Collins

Homework Book, Foundation 1
Delivering the Edexcel Specification

NEW GCSE MATHS
Edexcel Modular
Fully supports the 2010 GCSE Specification

D1427590

Brian Speed • Keith Gordon • Kevin Evans • Trevor Senior

CONTENTS

CORE

INTRODUCTION

Welcome to Collins New GCSE Maths for Edexcel Modular Foundation Homework Book 1. This book follows the structure of the Edexcel Modular Foundation Student Book and provides homework questions to cover topics in Unit 1.

Colour-coded grades

Know what target grade you are working at and track your progress with the colour-coded grade panels at the side of the page.

Use of calculators

Questions when you could use a calculator are marked with a  icon.

Examples

Recap on methods you need by reading through the examples before starting the homework exercises.

Functional maths

Practise functional maths skills to see how people use maths in everyday life. Look out for practice questions marked (FM).

There are also extra functional maths and problem-solving activities at the end of every chapter to build and apply your skills.

New Assessment Objectives

Practise new parts of the curriculum (Assessment Objectives AO2 and AO3) with questions that assess your understanding marked (AU) and questions that test if you can solve problems marked (PS). You will also practise some questions that involve several steps and where you have to choose which method to use; these also test AO2. There are also plenty of straightforward questions (AO1) that test if you can do the maths.

Student Book CD-ROM

Remind yourself of the work covered in class with the Student Book in electronic form on the CD-ROM. Insert the CD into your machine and choose the chapter you need.

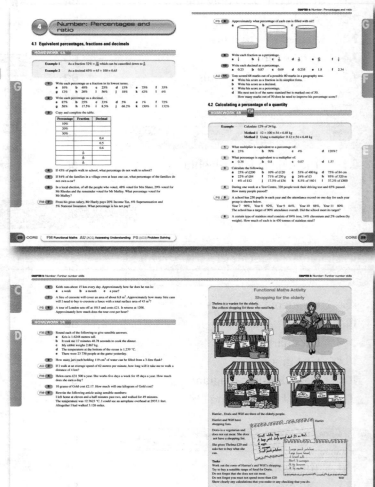

1 Number: Basic number

1.1 Adding with grids

1 Find the row and column sums of each of these grids.

a
2	5	1	☐
8	3	4	☐
9	7	6	☐
☐	☐	☐	☐

b
3	0	1	☐
7	5	6	☐
8	4	2	☐
☐	☐	☐	☐

c
0	3	5	☐
9	6	8	☐
2	7	1	☐
☐	☐	☐	☐

d
3	1	8	☐
6	2	0	☐
7	9	4	☐
☐	☐	☐	☐

e
8	1	3	☐
6	4	5	☐
2	7	0	☐
☐	☐	☐	☐

2 **a** Jack has six pens, four pencils and two rulers. How many items has he altogether?

b Hannah has an aquarium. The fish she has are four guppies, three neon tetras and five goldfish.
How many fish does she have altogether?

AU 3 Here is a list of numbers.

1   4   5   8   9   10

a From the list, write down **two** numbers that add up to 17.
b From the list, work out the largest total that can be made using **three** numbers.
c From the list, work out the largest **even** total that can be made using **three** numbers.

4 Asif has a £10 note, two £5 notes, a £2 coin and four £1 coins. His bus fare home is £1.80. Can he afford a shirt costing £25 and still catch the bus?

5 Find the numbers missing from each of these grids. Remember: the numbers missing from each grid must be chosen from 0 to 9 without any repeats.

a
1	5	☐	10
8	☐	7	17
☐	9	6	18
12	16	17	45

b
4	☐	5	10
☐	6	☐	8
7	8	☐	18
13	15	8	36

c
☐	☐	☐	14
0	9	8	17
2	☐	5	10
9	18	14	☐

d
4	☐	☐	12
☐	0	5	12
☐	8	3	☐
12	☐	10	36

e
☐	☐	3	10
☐	0	☐	☐
8	2	☐	15
16	☐	12	☐

PS 6 Find **three different** numbers that add up to 23.

PS 7 Find **two different odd** numbers that add up to 28.

1.2 Multiplication tables check

HOMEWORK 1B

1 Write down the answer to each of the following without looking at a multiplication square.

a	4×3	**b**	7×4	**c**	6×5	**d**	3×8	**e**	8×6
f	9×4	**g**	5×9	**h**	9×7	**i**	8×8	**j**	9×8
k	7×6	**l**	7×7	**m**	4×6	**n**	8×7	**o**	5×5

2 Jane works for 7 hours and is paid £9 an hour. How much is she paid?

3 Write down the answer to each of the following without looking at a multiplication square.

a	$14 \div 2$	**b**	$28 \div 4$	**c**	$24 \div 6$	**d**	$20 \div 5$	**e**	$18 \div 3$
f	$35 \div 5$	**g**	$27 \div 3$	**h**	$32 \div 4$	**i**	$24 \div 8$	**j**	$21 \div 7$
k	$42 \div 6$	**l**	$40 \div 8$	**m**	$18 \div 9$	**n**	$49 \div 7$	**o**	$48 \div 6$

4 Phil works for 6 hours and is paid £36. How much is he paid for each hour?

5 Write down the answer to each of the following. Look carefully at the signs, because they are a mixture of $+$, $-$, $\times$ and $\div$.

a	$8 + 5$	**b**	$20 - 6$	**c**	4×5	**d**	$16 \div 4$	**e**	$14 - 8$
f	$15 \div 3$	**g**	$16 + 8$	**h**	5×7	**i**	$16 + 5$	**j**	$36 \div 6$
k	$17 - 8$	**l**	9×3	**m**	$42 \div 7$	**n**	6×9	**o**	$21 - 6$

FM 6 Fatima works for 2 hours and is paid £12 an hour. Andy works for 3 hours and is paid £9 an hour. Who is paid the most?

AU 7 Here are four single-digit number cards.

2		8		4		5

The cards are used for making calculations.
Complete the following.

a $51 + \ldots \ldots \ldots = 135$

b $95 - \ldots \ldots \ldots = 50$

c $\ldots \ldots 7 + \ldots \ldots \ldots = 332$

8 Write down the answer to each of the following.

a	4×10	**b**	7×10	**c**	9×10	**d**	11×10	**e**	3×100
f	5×100	**g**	24×100	**h**	45×100	**i**	$80 \div 10$	**j**	$130 \div 10$
k	$510 \div 10$	**l**	$1000 \div 10$	**m**	$700 \div 100$	**n**	$900 \div 100$	**o**	$1200 \div 100$

PS 9 **a** Two consecutive numbers, when added together, give an answer of 21. What is the even number?

b Two whole numbers, when divided, give an answer of 15. One of the numbers is 4. What is the other number?

1.3 Order of operations and BIDMAS/BODMAS

HOMEWORK 1C

1 Work out each of these.
a $3 \times 4 + 7 =$ b $8 + 2 \times 4 =$ c $12 \div 3 + 4 =$ d $10 - 8 \div 2 =$
e $7 + 2 - 3 =$ f $5 \times 4 - 8 =$ g $9 + 10 \div 5 =$ h $11 - 9 \div 1 =$
i $12 \div 1 - 6 =$ j $4 + 4 \times 4 =$ k $10 \div 2 + 8 =$ l $6 \times 3 - 5 =$

2 Work out each of these. Remember: first work out the bracket.
a $3 \times (2 + 4) =$ b $12 \div (4 + 2) =$ c $(4 + 6) \div 5 =$
d $(10 - 6) + 5 =$ e $3 \times (9 \div 3) =$ f $5 + (4 \times 2) =$
g $(5 + 3) \div 2 =$ h $(5 \div 1) \times 4 =$ i $(7 - 4) \times (1 + 4) =$
j $(7 + 5) \div (6 - 3) =$ k $(8 - 2) \div (2 + 1) =$ l $15 \div (15 - 12) =$

3 Copy each of these and then put in brackets to make each sum true.
a $4 \times 5 - 1 = 16$ b $8 \div 2 + 4 = 8$ c $8 - 3 \times 4 = 20$ d $12 - 5 \times 2 = 2$
e $3 \times 3 + 2 = 15$ f $12 \div 2 + 1 = 4$ g $9 \times 6 \div 3 = 18$ h $20 - 8 + 5 = 7$
i $6 + 4 \div 2 = 5$ j $16 \div 4 \div 2 = 8$ k $20 \div 2 + 2 = 12$ l $5 \times 3 - 5 = 10$

AU 4 Jo says that $8 - 3 \times 2$ is equal to 10.
Show that Jo is wrong.

5 Put any of $+, -, \times, \div$ or ( ) in each sum to make it true.
a $2 \quad 5 \quad 10 = 0$ b $10 \quad 2 \quad 5 = 1$ c $10 \quad 5 \quad 2 = 3$ d $10 \quad 2 \quad 5 = 4$
e $10 \quad 5 \quad 2 = 7$ f $5 \quad 10 \quad 2 = 10$ g $10 \quad 5 \quad 2 = 13$ h $5 \quad 10 \quad 2 = 17$
i $10 \quad 2 \quad 5 = 20$ j $5 \quad 10 \quad 2 = 25$ k $2 \quad 2 \quad 2 = 2$

AU 6 Amanda worked out $3 + 4 \times 5$ and got the answer 35. Andrew worked out $3 + 4 \times 5$ and got the answer 23. Explain why they got different answers.

AU 7 You have to explain to someone how to work out this calculation.
$7 + 2 \times 6$
Write down what you would say.

PS 8 Here is a list of numbers, some symbols and one pair of brackets.
$2 \quad 5 \quad 6 \quad 42 \quad + \quad \times \quad = \quad (\quad)$
Use all of them to make a correct calculation.

PS 9 Here is a list of numbers, some symbols and one pair of brackets.
$1 \quad 3 \quad 5 \quad 8 \quad - \quad \div \quad = \quad (\quad)$
Use all of them to make a correct calculation.

FM 10 Jon has a piece of wood that is 8 metres long.

He wants to use his calculator to work out how much is left when he cuts off three pieces, each of length 1.2 metres.

Which calculation would give him the correct answer?
i $8 - 1.2 - 1.2 - 1.2$
ii $8 - 1.2 \times 3$
iii $8 - 1.2 + 1.2 + 1.2$

1.4 Rounding

HOMEWORK 1D

1 Round each of these numbers to the nearest 10.
   a 34 **b** 67 **c** 23 **d** 49 **e** 55
   f 11 **g** 95 **h** 123 **i** 109 **j** 125

2 Round each of these numbers to the nearest 100.
   a 231 **b** 389 **c** 410 **d** 777 **e** 850
   f 117 **g** 585 **h** 250 **i** 975 **j** 1245

3 Round each of these numbers to the nearest 1000.
   a 2176 **b** 3800 **c** 6760 **d** 4455 **e** 1204
   f 6782 **g** 5500 **h** 8808 **i** 1500 **j** 9999

4 The selling prices of five houses in a village are as follows:

FOR SALE	FOR SALE	FOR SALE	FOR SALE	FOR SALE
£8400	**£12 900**	**£45 300**	**£75 550**	**£99 500**

Give the prices to the nearest thousand pounds.

5 Give these bus journey times to the nearest 5 minutes.
   a 16 minutes **b** 28 minutes **c** 34 minutes **d** 42 minutes
   e $23\frac{1}{2}$ minutes **f** $17\frac{1}{2}$ minutes

6 Mark knows that he has £240 in his savings account to the nearest ten pounds.
   a What is the smallest amount that he could have?
   b What is the greatest amount that he could have?

7 The size of a crowd at an open-air pop festival was reported to be 8000 to the nearest thousand.
   a What is the lowest number that the crowd could be?
   b What is the largest number that the crowd could be?

FM 8 An estate agent values houses and rounds the value to the nearest £1000 for an advert and then subtracts £5.
   A house is valued at £89 600
   How much does the advert say?

PS 9 Mutasem and Ruba are playing a game with whole numbers.
   a Mutasem says, "I am thinking of a number. Rounded to the nearest 10, it is 270. What is the biggest number I could be thinking of?"
   b Ruba says, "I am thinking of a different number. Rounded to the nearest 100, it is 300. It is less than 270. How many possible answers are there?"

PS 10 The number of fish in a pond is 130 to the nearest 10.
   The number of frogs in the pond is 90 to the nearest 10.
   Show how there could be 230 fish and frogs altogether in the pond.

1.5 Adding and subtracting numbers with up to four digits

HOMEWORK 1E

1 Copy and work out each of these additions.

a	75	**b**	245	**c**	307	**d**	4158	**e**	4289
	+ 23		+ 156		+ 293		+ 3951		532
									+ 96

2 Complete each of these additions.

a 25 + 89 + 12 **b** 211 + 385 + 46 **c** 125 + 88 + 720

d 478 + 207 + 300 **e** 1275 + 3245 + 524

3 Copy and complete each of these subtractions.

a	354	**b**	651	**c**	785	**d**	450	**e**	5421
	−120		−128		−207		−178		−2568

4 Complete each of these subtractions.

a 386 − 296 **b** 709 − 518 **c** 452 − 386

d 800 − 258 **e** 7208 − 1564

FM 5 The train from Brighton to London takes 68 minutes.

The train from London to Birmingham takes 85 minutes.

a How long does it take to travel from Brighton to London and then London to Birmingham altogether, if there is a 30-minute wait between trains in London?

b How much longer does the train take from London to Birmingham than from Brighton to London?

AU 6 Michael is checking the addition of two numbers.

His answer is 917.

One of the numbers is 482.

What should the other number be?

PS 7 Copy each of these and fill in the missing digits.

a	4 5	**b**	□7	**c**	3□4	**d**	□□□
	+3□		+4□		+2 8 6		+ 2 8 7
	□7		9 2		□4□		5 5 5

PS 8 Copy each of these and fill in the missing digits.

a	7 5	**b**	3 2□	**c**	5 8 3	**d**	□□□
	− 1□		−1 □ 4		−□□□		− 2 4 8
	□3		1 8 2		1 3 5		3 7 4

1.6 Multiplying and dividing by single-digit numbers

HOMEWORK 1F

1 Copy and work out each of the following.

a	24	**b**	38	**c**	124	**d**	408	**e**	359
	× 3		× 4		× 5		× 6		× 8

2 Calculate each of these multiplications.

  a 21×5 **b** 37×7 **c** 203×9 **d** 4×876 **e** 6×3214

3 Alex, Peter and Theo are footballers.

The manager of their club has offered them a bonus of £5 for every goal they score.

  Alex scores 15 goals

  Peter scores 12 goals

  Theo scores 20 goals

  a How many goals do they score altogether?

  b How much bonus does each footballer receive?

4 Calculate each of these divisions.

  a $684 \div 2$ **b** $525 \div 3$ **c** $804 \div 4$ **d** $7260 \div 5$ **e** $2560 \div 8$

PS FM 5 The footballers in question 3 are given a goals target for the following season.

The manager wants the total of the goals target to add up to 55, with Theo having the highest target and Peter the lowest.

Show how this can be done.

	Goals Target
Alex	
Peter	
Theo	
	Total = 55

6 By doing a suitable multiplication, answer each of these questions.

  a How many people could seven 55-seater coaches hold?

  b Adam buys seven postcards at 23p each. How much does he spend in pounds?

  c Nails are packed in boxes of 144. How many nails are there in five boxes?

  d Eight people book a holiday, costing £284 each. What is the total cost?

  e How many yards are there in six miles if there are 1760 yards in a mile?

7 By doing a suitable division, answer each of these questions.

  a In a school there are 288 students in eight forms in Year 10. If there are the same number of students in each form, how many students are there in each one?

  b Phil jogs seven miles every morning. How many days will it take him to cover a total distance of 441 miles?

  c In a supermarket, cans of cola are sold in packs of six. If there are 750 cans on the shelf, how many packs are there?

  d Sandra's wages for a month were £2060. Assuming there are four weeks in a month, how much does she earn in a week?

  e Tickets for a charity disco were sold at £5 each. How many people bought tickets if the total sales were £1710?

Functional Maths Activity

Hairdressing

The prices at the hairdressers change according to the amount of experience that the hairdresser has. Stylists charge the least and senior technicians charge the most.

Cut & finishing prices:	Stylists	Senior stylists	Senior Technicians
Ladies' cut & blow dry	£23	£31	£36
Ladies' restyle	£26	£33	£38
Ladies' blow dry	£15	£17	£18
Ladies' hair up	–	£25	£25
Men's cut & finish	£12	£18	£24
Boys	£8	£10	–

Technical prices:	Stylists	Senior stylists	Senior Technicians
Highlights Foil (full head)	£51	£60	£67
Highlights Foil (half head)	£39	£45	£52
Highlights Foil (partial from...)	£15	£20	£20
Tinting (full head)	£39	£43	£50
Tinting (re-growth)	£34	£38	£42
Permanent Waving (from...)	£45	£50	£57

Special Offer

Wednesdays only

Pensioners £5 off when spending £20 or less

£10 off when spending more than £20

The Cutts family visit the hairdressers on a Wednesday.

Work out how much it will cost for each person.

I want the cheapest cut I can get.

I always have full head tints and a cut and blow dry.

I'm having half-head highlights with a restyle, so I want a senior stylist.

I want my hair up.

I don't care who cuts my hair.

How much do they save by going on Wednesday?

Do the same for your own family by deciding which haircuts to give them:
- Choose at least four members of your family.
- Decide which day of the week to go to the hairdressers.
- Do not make the same choices as the Cutts family.
- At least one of your choices should use the senior technician.

Number: Fractions

2.1 Recognise a fraction of a shape

HOMEWORK 2A

1 What fraction is shaded in each of these diagrams?

a b c

d  e 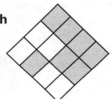 f

g h

2 Draw diagrams as in Question **1** to show these fractions.

a $\frac{1}{3}$ b $\frac{3}{5}$ c $\frac{7}{10}$ d $\frac{5}{8}$ e $\frac{7}{9}$

f $\frac{3}{7}$ g $\frac{5}{12}$ h $\frac{7}{15}$

PS 3 Look again at the diagrams in Question **1**.

For each of the following pairs of diagrams, decide which diagram has the greater proportion shaded.

i   b and e

ii a and d

iii c and g

FM 4 A dressmaker cuts a piece of cloth into two equal parts. She then cuts one of the parts into two equal parts.

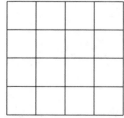

Make two copies of the grid above and show two different ways that she can do this by shading the smaller area in each case.

FM Functional Maths   **AU** (AO2) Assessing Understanding   **PS** (AO3) Problem Solving

AU 5 Here are three grids A, B and C.

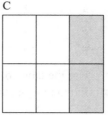

Give a reason why grid B could be the odd one out.

2.2 Recognise equivalent fractions, using diagrams

HOMEWORK 2B

1 Copy the diagram and use it to write down each of these fractions as tenths.

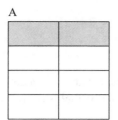

a $\frac{1}{2}$ **b** $\frac{1}{5}$ **c** $\frac{2}{5}$ **d** $\frac{3}{5}$ **e** $\frac{4}{5}$

2 Use your answers to Question **1** to write down the answer to each of the following. Each answer will be expressed in tenths.

a $\frac{1}{2} + \frac{1}{5}$ **b** $\frac{1}{2} + \frac{3}{10}$ **c** $\frac{2}{5} + \frac{1}{10}$ **d** $\frac{1}{5} + \frac{7}{10}$ **e** $\frac{1}{2} - \frac{2}{5}$ **f** $\frac{9}{10} - \frac{3}{5}$

3 Copy the diagram and use it to write down each of these fractions as twelfths.

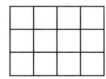

a $\frac{1}{2}$ **b** $\frac{1}{4}$ **c** $\frac{1}{3}$ **d** $\frac{3}{4}$ **e** $\frac{2}{3}$

4 Use your answers to Question **3** to write down the answer to each of the following. Each answer will be expressed in twelfths.

a $\frac{1}{2} + \frac{1}{3}$ **b** $\frac{1}{4} + \frac{1}{3}$ **c** $\frac{2}{3} + \frac{1}{4}$ **d** $\frac{1}{4} + \frac{7}{12}$ **e** $\frac{5}{12} + \frac{1}{2}$

f $\frac{2}{3} - \frac{1}{2}$ **g** $\frac{3}{4} - \frac{1}{3}$ **h** $\frac{1}{2} - \frac{1}{12}$ **i** $\frac{3}{4} - \frac{5}{12}$ **j** $\frac{2}{3} - \frac{7}{12}$

PS 5 Use the grid to decide which of the following fractions is the odd one out.

$\frac{5}{6}$ $\frac{3}{4}$ $\frac{10}{12}$

AU 6 Here is a multiplication table.

×	2	3	4	5
3	6	9	12	15
5	10	15	20	25

a Use the table to write down three more fractions that are equivalent to $\frac{3}{5}$.

b Extend this method to write down a different fraction that is equivalent to $\frac{3}{5}$.

FM 7 a Use a copy of the grid in Question **5** to write down the answer to $\frac{1}{4} + \frac{1}{12}$.

b A man has £120.

He gives one-quarter of it to his wife and one-twelfth to his son.

He gives the rest to his daughter. She see a dress advertised at half price. The normal price is £150. Will she be able to afford it?

2.3 Equivalent fractions and simplifying fractions by cancelling

HOMEWORK 2C

Example 1 Show $\frac{2}{3}$ is equivalent to $\frac{8}{12}$.

$$\frac{2}{3} \rightarrow \frac{\times 4}{\times 4} = \frac{8}{12}$$

Example 2 Cancel down $\frac{15}{20}$ to its simplest form.

$$\frac{15}{20} \rightarrow \frac{\div 5}{\div 5} = \frac{3}{4}$$

Example 3 Put these fractions in order with the smallest first: $\frac{5}{6}$ $\frac{2}{3}$ $\frac{3}{4}$.

By equivalent fractions $\frac{5}{6} = \frac{10}{12}$ $\frac{2}{3} = \frac{8}{12}$ $\frac{3}{4} = \frac{9}{12}$

Putting in order: $\frac{2}{3}$ $\frac{3}{4}$ $\frac{5}{6}$

1 Copy and complete each of these statements.

a $\frac{1}{4} \rightarrow \frac{\times 3}{\times 3} \frac{}{12}$ b $\frac{3}{5} \rightarrow \frac{\times 4}{\times 4} = \frac{}{20}$ c $\frac{5}{8} \rightarrow \frac{\times 2}{\times 2} \frac{}{16}$ d $\frac{4}{7} \rightarrow \frac{\times 3}{\times 3} = \frac{12}{}$

e $\frac{2}{3} \rightarrow \frac{\times}{\times 5} = \frac{}{15}$ f $\frac{5}{9} \rightarrow \frac{\times}{\times 2} = \frac{}{18}$ g $\frac{6}{7} \rightarrow \frac{\times}{\times} = \frac{}{35}$ h $\frac{1}{10} \rightarrow \frac{\times}{\times} = \frac{}{40}$

2 Copy and complete each of these statements.

a $\frac{1}{4} = \frac{2}{8} = \frac{}{12} = \frac{4}{} = \frac{}{20} = \frac{6}{24}$ b $\frac{2}{3} = \frac{4}{6} = \frac{}{} = \frac{}{12} = \frac{10}{} = \frac{12}{18}$

c $\frac{4}{5} = \frac{}{10} = \frac{12}{} = \frac{}{20} = \frac{}{} = \frac{}{30}$ d $\frac{3}{10} = \frac{}{} = \frac{}{30} = \frac{}{} = \frac{}{50} = \frac{18}{}$

3 Copy and complete each of these statements.

a $\frac{6}{8} = \frac{6 \div 2}{8 \div 2} = \frac{}{}$ b $\frac{9}{12} = \frac{9 \div 3}{12 \div 3} = \frac{}{}$

c $\frac{15}{25} = \frac{15 \div 5}{25 \div} = \frac{}{}$ d $\frac{20}{70} = \frac{20 \div 10}{70 \div} = \frac{}{}$

4 A woman uses her garage to store different items.
She divides the area up into four equal sections.
In one section she put all her items for recycling. In two other sections she stores wine.
The fourth section is empty.
  a What fraction of the garage has wine in it?
     Give your answer in its simplest form.
  b What fraction of the garage is not empty?

5 Cancel each of these fractions to its simplest form.
  a $\frac{4}{10}$ **b** $\frac{3}{12}$ **c** $\frac{5}{25}$ **d** $\frac{6}{15}$ **e** $\frac{8}{12}$
  f $\frac{10}{30}$ **g** $\frac{12}{20}$ **h** $\frac{16}{24}$ **i** $\frac{30}{50}$ **j** $\frac{42}{49}$

6 Put the following fractions in order with the **smallest** first.
  a $\frac{1}{3}, \frac{1}{2}, \frac{1}{4}$ **b** $\frac{3}{4}, \frac{3}{8}, \frac{1}{2}$ **c** $\frac{5}{6}, \frac{2}{3}, \frac{7}{12}$ **d** $\frac{2}{5}, \frac{3}{10}, \frac{1}{4}$

AU 7 Here are four unit fractions.
     $\frac{1}{2}$ $\frac{1}{3}$ $\frac{1}{4}$ $\frac{1}{5}$
  a Which two of these fractions have a sum of $\frac{5}{6}$?
     Show clearly how you work out your answer.
  b Which fraction is the smallest?
     Explain your answer.

2.4 Finding a fraction of a quantity

HOMEWORK 2D

1 Calculate each of these.
  a $\frac{1}{2}$ of 20   **b** $\frac{1}{3}$ of 36   **c** $\frac{1}{4}$ of 24
  d $\frac{3}{4}$ of 40   **e** $\frac{2}{3}$ of 15   **f** $\frac{1}{5}$ of 30
  g $\frac{3}{8}$ of 16   **h** $\frac{7}{10}$ of 50

2 Calculate each of these quantities.
  a $\frac{1}{4}$ of £800   **b** $\frac{2}{3}$ of 60 kilograms   **c** $\frac{3}{4}$ of 200 metres
  d $\frac{3}{8}$ of 48 gallons   **e** $\frac{4}{5}$ of 30 minutes   **f** $\frac{7}{10}$ of 120 miles

3 In each case, find out which is the smaller number.
  a $\frac{1}{4}$ of 60 or $\frac{1}{2}$ of 40   **b** $\frac{1}{3}$ of 36 or $\frac{1}{5}$ of 50
  c $\frac{2}{3}$ of 15 or $\frac{3}{4}$ of 12   **d** $\frac{5}{8}$ of 72 or $\frac{5}{6}$ of 60

4 $\frac{5}{9}$ of a class of 36 students are girls. How many boys are there in the class?

FM 5 Mrs Wilson puts $\frac{3}{20}$ of her weekly wage into a pension scheme. How much does she put into the scheme if her wage one week is £320?

6 Mitchell spent one third of a day sleeping and one quarter at school. How many hours are left for doing other things?

7 A bush is 40 cm tall when planted in spring. Its height increases by $\frac{3}{10}$ during the summer.
  a Find $\frac{3}{10}$ of 40 cm.
  b Find the height of the bush at the end of the summer.

FM 8 A travel agent has this sign in its window. Marion books a holiday for her family which would normally cost £800.

1/5 OFF

all holiday prices for next year if booked before December

  a   How much does she save?

  b   How much does she pay for the holiday after the reduction?

PS 9 A chair is advertised at Top Sofa for £800, including a free footstool, but with an offer of one-quarter off.

The same chair is advertised at Comfy Seats for £480 and the footstool costs £130 extra. Which shop is cheaper for buying both the chair and the footstool?

2.5 One quantity as a fraction of another

HOMEWORK 2E

Example   Write £5 as a fraction of £20.

$\frac{5}{20} = \frac{1}{4}$ (Cancel down)

1   Write the first quantity as a fraction of the second.
  a   £2, £8 **b**   9 cm, 12 cm **c**   18 miles, 30 miles
  d   200 g, 350 g **e**   20 seconds, 1 minute **f**   25p, £2

2   During a one-hour TV programme, 10 minutes were devoted to adverts. What fraction of the time was given to adverts?

AU 3   On an 80-mile car journey, 50 miles were driven on a motorway. What fraction of the journey was not driven on a motorway?

4   In a class, 24 students were right-handed and 6 students were left-handed. What fraction of the class were:
  a   right-handed **b**   left-handed?

FM 5   Mark earns £120 and saves £40 of it.
Bev earns £150 and saves £60 of it.
Who is saving the greater proportion of their earnings?

FM 6   In a test, Kevin scores 7 out of 10 and Sally scores 15 out of 20. Which is the better mark? Explain your answer.

2.6 Arithmetic with fractions

HOMEWORK 2F

Example 1   Express 0.32 as a fraction.

$0.32 = \frac{32}{100}$. This cancels down to $\frac{8}{25}$.

Example 2   Express $\frac{3}{8}$ as a decimal. $\frac{3}{8} = 3 \div 8 = 0.375$.

$$\begin{array}{r} 0.375 \\ 8\overline{)3.^30^60^40} \end{array}$$ Notice how the extra zeros have been added.

1 Change each of these decimals to fractions, cancelling down where possible.

a 0.3 **b** 0.8 **c** 0.9 **d** 0.07 **e** 0.08

f 0.15 **g** 0.75 **h** 0.48 **i** 0. 32 **j** 0.27

2 Change each of these fractions to decimals.

a $\frac{1}{4}$ **b** $\frac{2}{5}$ **c** $\frac{7}{10}$ **d** $\frac{9}{20}$ **e** $\frac{7}{8}$

3 Put each of the following sets of numbers in order with the smallest first. It is easier to change the fractions into decimals first.

a 0.3, 0.2, $\frac{2}{5}$ **b** $\frac{7}{10}$, 0.8, 0.6 **c** 0.4, $\frac{1}{4}$, 0.2

d $\frac{3}{10}$, 0.32, 0.29 **e** 0.81, $\frac{4}{5}$, 0.78

FM 4 Two shops are advertising the same skirts.

Which shop has the better offer?
Give a reason for your answer.

PS 5 Which is bigger: $\frac{5}{6}$ or 0.8?
Show your working.

PS 6 Which is smaller: $\frac{1}{3}$ or 0.3?
Show your working.

PS 7 Complete this statement with a fraction.
$\frac{1}{3}$ of 45 is the same as ...... of 30.

HOMEWORK 2G

1 Evaluate the following.

a $\frac{1}{2}+\frac{1}{5}$ **b** $\frac{1}{2}+\frac{1}{3}$ **c** $\frac{1}{3}+\frac{1}{10}$ **d** $\frac{3}{8}+\frac{1}{3}$

e $\frac{3}{4}+\frac{1}{5}$ **f** $\frac{1}{3}+\frac{2}{5}$ **g** $\frac{3}{5}+\frac{3}{8}$ **h** $\frac{1}{2}+\frac{2}{5}$

2 Evaluate the following.

a $\frac{1}{2}+\frac{1}{4}$ **b** $\frac{1}{3}+\frac{1}{6}$ **c** $\frac{3}{5}+\frac{1}{10}$ **d** $\frac{5}{8}+\frac{1}{4}$

3 Evaluate the following.

a $\frac{7}{8}-\frac{3}{4}$ **b** $\frac{4}{5}-\frac{1}{2}$ **c** $\frac{2}{3}-\frac{1}{5}$ **d** $\frac{3}{4}-\frac{2}{5}$

4 Evaluate the following.

a $\frac{5}{8}+\frac{3}{4}$ **b** $\frac{1}{2}+\frac{3}{5}$ **c** $\frac{5}{6}+\frac{1}{4}$ **d** $\frac{2}{3}+\frac{3}{4}$

5 Evaluate the following.

a $2\frac{1}{3}+1\frac{1}{4}$ **b** $3\frac{7}{10}+2\frac{3}{4}$ **c** $5\frac{3}{8}-2\frac{1}{3}$ **d** $4\frac{2}{5}-2\frac{5}{6}$

6 At a football club half of the players are English, a quarter are Scottish and one-sixth are Italian. The rest are Irish. What fraction of players at the club are Irish?

7 On a firm's coach trip, half the people were employees, two-fifths were partners of the employees. The rest were children. What fraction of the people were children?

AU 8 Five-eighths of the 35 000 crowd were male. How many females were in the crowd?

9 What is four-fifths of sixty-five added to five-sixths of fifty-four?

10 Find two fractions with a sum of $\frac{11}{12}$.

HOMEWORK 2H

1 Evaluate the following, leaving your answer in its simplest form.

a $\frac{1}{2} \times \frac{2}{3}$ b $\frac{3}{4} \times \frac{2}{5}$ c $\frac{3}{5} \times \frac{1}{2}$ d $\frac{3}{7} \times \frac{2}{3}$ e $\frac{2}{3} \times \frac{5}{6}$

f $\frac{1}{3} \times \frac{3}{5}$ g $\frac{2}{3} \times \frac{7}{10}$ h $\frac{3}{8} \times \frac{2}{5}$ i $\frac{4}{9} \times \frac{3}{8}$ j $\frac{4}{5} \times \frac{7}{16}$

2 Kris walked three-quarters of the way along Carterknowle Road, which is 3 km long. How far did Kris walk?

AU 3 Jean ate one-fifth of a cake, Les ate a half of what was left. Nick ate the rest. What fraction of the cake did Nick eat?

FM 4 a Billie made a cast that weighed five and three quarter kilograms. Four-fifths of this weight is water. What is the weight of the water in Billie's cast?

b 1 litre of water weighs 1 kg. A jug holds 2 litres of water. How many times does Billie need to fill the jug to have enough water to make the cast.

5 Evaluate the following, leaving your answer as a mixed number where possible.

a $1\frac{1}{3} \times 2\frac{1}{4}$ b $1\frac{3}{4} \times 1\frac{1}{3}$ c $2\frac{1}{2} \times \frac{4}{5}$ d $1\frac{2}{3} \times 1\frac{3}{10}$

e $3\frac{1}{4} \times 1\frac{1}{5}$ f $2\frac{2}{3} \times 1\frac{3}{4}$ g $3\frac{1}{2} \times 1\frac{1}{6}$ h $7\frac{1}{2} \times 1\frac{3}{5}$

AU 6 Which is the smaller, $\frac{3}{4}$ of $5\frac{1}{3}$ or $\frac{2}{3}$ of $4\frac{2}{5}$?

FM PS 7 I estimate that I need 60 litres of lemonade for a party.

I buy 24 bottles, which each contain $2\frac{3}{4}$ litres.

Have I bought enough lemonade for the party?

HOMEWORK 2I

1 Evaluate the following, leaving your answer as a mixed number where possible.

a $\frac{1}{5} \div \frac{1}{3}$ b $\frac{3}{5} \div \frac{3}{8}$ c $\frac{4}{5} \div \frac{2}{3}$ d $\frac{4}{7} \div \frac{8}{9}$

e $4 \div 1\frac{1}{2}$ f $5 \div 3\frac{2}{3}$ g $8 \div 1\frac{3}{4}$ h $6 \div 1\frac{1}{4}$

i $5\frac{1}{2} \div 1\frac{1}{3}$ j $7\frac{1}{2} \div 2\frac{2}{3}$ k $1\frac{1}{2} \div 1\frac{1}{5}$ l $3\frac{1}{5} \div 3\frac{3}{4}$

FM 2 A pet shop has thirty-six kilograms of hamster food. Tom, who owns the shop, wants to pack this into bags, each containing three-quarters of a kilogram. How many bags can he make in this way?

AU 3 Bob is putting a fence down the side of his garden, it is to be 20 metres long. The fence comes in sections; each one is one and one-third of a metre long. How many sections will Bob need to put the fence all the way down the one side of his garden?

4 An African Bullfrog can jump a distance of $1\frac{1}{4}$ metres in one hop. How many hops would it take an African Bullfrog to hop a distance of 100 metres?

Functional Maths Activity

Organising the school timetable

In a school there are **180** students in year 10 and you want to organise them in to classes for different option subjects.

The Heads of Department give you the following facts:

Languages (Spanish and French)

All students take one language, except for one group of 26 students who go out on a work placement.

Spanish

Half of all students do Spanish.
They are all taught at the same time.
There are three equal-sized classes.

French

There are three classes.
The smallest class has one-quarter of the students taking French.
The other two classes have the same number of students in each.

Geography

Two-thirds of all students in the year group take geography.
There are five equal-sized classes.

History

Two-fifths of all students take history.
One class has 21 students and one class has 18 students.
There are two other classes. Two-thirds of the remaining students are in one class.

Copy and complete the table for the classes. Remember that there are 180 students in the year group. Choose your own rules for PE.

Decide the number of classes and work out the number in each class so that no PE class exceeds 30 students.

Subject	Number in each class		
Spanish			
French			
Geography			
History			
PE			

Now copy and complete this table to give the fraction of the year group in each class.

Subject	Fraction of whole year group in each class		
Spanish			
French			
Geography			
History			
PE			

3 Number: Negative numbers

3.1 Introduction to negative numbers

HOMEWORK 3A

1 Write down the temperatures for each thermometer.

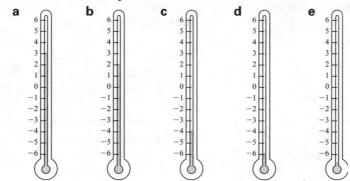

a   b   c   d   e

2 Look at this map showing average temperatures in the capital cities of England, Scotland and Wales.

a How much colder is it in Edinburgh than Cardiff?

b How much warmer is it in Cardiff than Edinburgh?

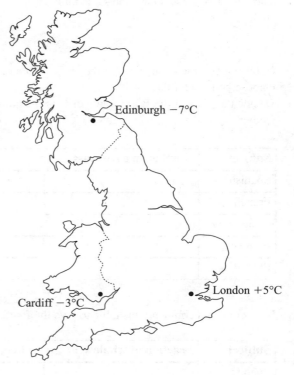

Edinburgh −7°C

Cardiff −3°C

London +5°C

FM Functional Maths   **AU** (AO2) Assessing Understanding   **PS** (AO3) Problem Solving

AU PS **3** The diagram shows the layout of a hotel and the lift numbers for each floor.

a How many floors is the third floor above the basement?

b How many floors is the lower ground floor below the first floor?

c How many floors is the basement below the second floor?

d The cellar basement is three floors below the basement. What number is used for the lift?

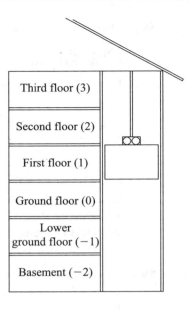

| Third floor (3) |
| Second floor (2) |
| First floor (1) |
| Ground floor (0) |
| Lower ground floor (−1) |
| Basement (−2) |

3.2 Everyday use of negative numbers

HOMEWORK 3B

Copy and complete each of the following.

1 If +£20 means a profit of twenty pounds, then .... means a loss of twenty pounds.

2 If −£10 means a loss of ten pounds, then +£10 means a .... of ten pounds.

3 If +500 m means 500 metres above sea level, then .... means 500 metres below sea level.

4 If −1000 m means one thousand metres below sea level, then +1000 m means one thousand metres .... sea level.

5 If +7°C means seven degrees above freezing point, then .... means seven degrees below freezing point.

6 If +1°C means 1°C above freezing point, then .... means 1°C below freezing point.

7 If −15°C means fifteen degrees below freezing point, then +15°C means fifteen degrees .... freezing point.

8 If −5000 miles means five thousand miles south of the equator, then +5000 miles means five thousand miles .... of the equator.

9 If a car moving forwards at 25 mph is represented by +25 mph, then a car moving backwards at 10 mph is represented by ....

10 In multi-storey car park, the sixth floor above ground level is represented by +6. So, the third floor below ground level is represented by ....

PS **11** A hotel has 11 levels numbered from basement (−1) to ninth floor (+9).
A man is in a lift on the third floor. He goes up four levels and then down to the basement.
How many levels does the lift travel while descending?

FM 12 Here is Faisal's bank statement.

Sproggit & Sylvester Bank

Customer ID: 5436t499
Statement no: 4565
Statement date: 29 SEP 2009

Date	Description	Debits	Credits	Balance
21 Sept 2009	Electricity bill 125.48...	£125.32	...	£250.80
24 Sept 2009	Cheque	...	£30.00	£155.48
29 Sept 2009	Water bill	£162.50	...	−£7.02

a How much does he owe the bank?
b On 30th September he pays in £10.
How much has he in the bank now?

AU 13 The temperature on three days in Quebec in Canada was −9°C, −8°C and −11°C.
a Put the temperatures in order starting with the coldest.
b What is the difference in temperature between the coldest and the warmest temperature?

3.3 The number line

HOMEWORK 3C

Use the number line to answer Questions **1** and **2**.

−7 −6 −5 −4 −3 −2 −1 0 1 2 3 4 5 6 7

negative   **positive**

1 Complete each of the following by putting a suitable number in the box.
a ☐ is smaller than 6   **b** ☐ is smaller than 2
c ☐ is smaller than −1   **d** ☐ is smaller than −6
e −4 is smaller than ☐   **f** −7 is smaller than ☐
g 5 is smaller than ☐   **h** 4 is smaller than ☐
i ☐ is smaller than 0   **j** −1 is smaller than ☐

2 Complete each of the following by putting a suitable number in the box.
a ☐ is bigger than −6   **b** ☐ is bigger than 4
c ☐ is bigger than −2   **d** ☐ is bigger than 0
e −3 is bigger than ☐   **f** −5 is bigger than ☐
g 1 is bigger than ☐   **h** −1 is bigger than ☐
i ☐ is bigger than −4   **j** 3 is bigger than ☐

3 In each case below, put the correct symbol, either < or >, in the box.
a 2 ☐ 6   **b** −1 ☐ −7
c −5 ☐ 1   **d** 5 ☐ 9
e −8 ☐ 2   **f** −14 ☐ −10
g −11 ☐ 0   **h** −9 ☐ −12
i 8 ☐ −3   **j** 0 ☐ −8

HINTS AND TIPS

Reminder: The inequality signs: < means 'is less than' and > means 'is greater than'.

PS 4 Copy these number lines and fill in the missing numbers on each line.

a

b

c

d

e

FM 5 Here are some temperatures.

    3°C   −8°C   −1°C   4°C

Copy and complete the following weather report using these temperatures.

Temperatures fell to ...., making Newcastle the coldest place in the country today, while in Yorkshire the temperatures were also below freezing at .... People on the south coast enjoyed warmer temperatures, with Eastbourne being the warmest at .... and Brighton only slightly lower at ....

AU 6 Here are some numbers.

    $-2\frac{3}{4}$ $+1\frac{1}{2}$ $-1\frac{1}{4}$

Copy the number line below and mark the numbers on it.

    −3   −2   −1   0   1   2   3

3.4 Arithmetic with negative numbers

HOMEWORK 3D

Example 1 $-3 + 5 = 2$

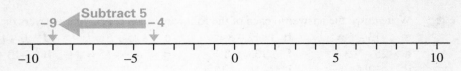

Example 2 $-4 - 5 = -9$

1 Use the number line to find the answer to each of the following.

a $-2 + 5 =$ **b** $-4 + 6 =$ **c** $-3 + 4 =$ **d** $-1 + 5 =$

e $-6 + 8 =$ **f** $-5 + 10 =$ **g** $-2 + 2 =$ **h** $-4 + 4 =$

i $4 - 5 =$ **j** $6 - 8 =$ **k** $3 - 7 =$ **l** $5 - 9 =$

m $-5 + 3 =$ **n** $-2 + 1 =$ **o** $-10 + 6 =$ **p** $-8 + 6 =$

q $-2 - 7 =$ **r** $-1 - 5 =$ **s** $-3 - 7 =$ **t** $-5 - 5 =$

2 Answer each of the following without the help of the number scale.

a $15 - 19 =$ **b** $3 - 17 =$ **c** $-2 - 10 =$ **d** $-12 + 7 =$

e $-15 + 9 =$ **f** $10 - 20 =$ **g** $-10 - 12 =$ **h** $-15 - 20 =$

i $23 - 30 =$

3 Work out each of the following.

a $1 + 3 - 5 =$ **b** $-4 + 8 - 2 =$ **c** $-6 + 3 - 4 =$ **d** $-3 - 5 + 4 =$

e $-1 - 1 + 5 =$ **f** $-7 + 5 + 8 =$ **g** $-3 - 4 + 7 =$ **h** $1 - 3 - 6 =$

i $-5 - 3 - 2 =$

4 Check your answers to questions 1 to 3 using a calculator.

AU 5 At noon the temperature in Glasgow was –3°C.

At 2 pm the temperature had risen by 2°C.

a What is the temperature at 2 pm?

b At 6 pm the temperature was 5 degrees lower than it was at 2 pm. What was the temperature at 6 pm?

PS 6 Here are five numbers.

    5   9   1   3   4

a Use two of the numbers to make a calculation with an answer of –5.

b Use three of the numbers to make a calculation with an answer of –11.

c Use four of the numbers to make a calculation with an answer of –10.

d Use all five of the numbers to make a calculation with an answer of –20.

HOMEWORK 3E

Example 1 $4 - (-2) = 4 + 2 = 6$

Example 2 $3 + (-5) = 3 - 5 = -2$

1 Answer each of the following. Check your answers on a calculator.

a $2 + (-5) =$ **b** $6 - (-3) =$ **c** $3 + (-5) =$ **d** $8 - (-2) =$

e $-6 + (-2) =$ **f** $-5 + (-2) =$ **g** $-2 - (-5) =$ **h** $-7 - (-1) =$

i $-2 - (-2) =$

2 Write down the answer to each of the following, then check your answers on a calculator.

a $-13 - 5 =$ **b** $-12 - 8 =$ **c** $-25 + 6 =$ **d** $6 - 14 =$

e $25 - -3 =$ **f** $13 - -8 =$ **g** $-4 + -15 =$ **h** $-13 + -7 =$

i $-12 + -9 =$ **j** $-16 + -12 =$

3 The temperature at midday was 5°C. Find the temperature at midnight if it fell by:

a 1°C   **b** 5°C   **c** 6°C   **d** 8°C   **e** 12°C.

4 What is the difference between the following temperatures?

    a 4°C and 6°C   **b** −2°C and 4°C   **c** −3°C and −6°C

5 Rewrite the following lists, putting the numbers in order of size, smallest first.

    a 2 −5 3 −6 −3 8 −1 1

    b 4 −8 5 −10 −5 0 6 −12

AU 6 Write down five addition and five subtraction calculations that give an answer of −5.

AU 7 You have the following cards.

5	3	2	−1	−3	−6

    a Which other card should you choose to make the answer to the following sum as large as possible? What is the answer?

    5 **+** ☐ **=** ☐

    b Which other card should you choose to make the answer to part **a** as small as possible? What is the answer?

    c Which other card should you choose to make the answer to the following sum as large as possible? What is the answer?

    5 **−** ☐ **=** ☐

    d Which other card should you choose to make the answer to part **c** as small as possible? What is the answer?

    e Which two cards should you choose to make the answer to an addition sum zero?

8 Two numbers have a sum of 8. One of the numbers is negative. The other number is odd. What could the numbers be? Give two different answers

HOMEWORK 3F

1 Find the next three numbers in each sequence.

    a 6, 4, 2, 0, …, …, …   **b** 8, 5, 2, −1, …, …, …

    c −20, −15, −10, −5, …, …, …   **d** 10, 9, 7, 4, …, …, …

    e −12, −10$\frac{1}{2}$, −9, −7$\frac{1}{2}$, …, …, …

FM 2 The deep freeze compartment in a refrigerator should be set at −14°C, but in error is set to −6°C. How does the temperature setting need to be changed so that it is correct?

FM 3 At 5am, the temperature on a thermometer outside Brian's house was −4°C. By midday, the temperature had risen by 10°C.

    a What was the temperature at midday?

    By midnight, the temperature had fallen to −9°C.

    b What was the fall in temperature from midday to midnight?

4 The table shows the recorded highest and lowest temperatures in five cities during one year.

	London	New York	Athens	Beijing	Nairobi
Highest temperature (°C)	30	28	36	31	29
Lowest temperature (°C)	−5	−8	5	−10	11

    a Which city had the highest temperature?

    b Which city had the largest difference in temperature and by how many degrees?

    c Which city had the smallest difference in temperature and by how many degrees?

5 This is a two-step function machine. Use the function machine to complete the table.

| Number in | ➡ | Add 2 | ➡ | Subtract 5 | ➡ | Number out |

Number in	Number out
10	
2	
−1	
	0
	−8

PS 6 Copy and complete this magic square. Write down the magic number.

1	−6	−1
−4		

Functional Maths Activity

The hotel lift

In a hotel, there are seven floors and one lift, as shown in the diagram opposite.

The manager of the hotel works out that the average time for a lift to move from one floor to the next is 15 seconds.

Reception is on the ground floor.

Bedrooms are on the first to fourth floors.

The gymnasium is in the lower basement.

Guests have complained that the lifts take too long to arrive.

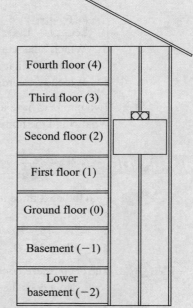

| Fourth floor (4) |
| Third floor (3) |
| Second floor (2) |
| First floor (1) |
| Ground floor (0) |
| Basement (−1) |
| Lower basement (−2) |

Task 1

Complete the table to work out the number of floors the lift has to travel (− for down and + for up).

The first few have been done for you.

		From						
		−2	**−1**	**0**	**1**	**2**	**3**	**4**
	−2	——	−1					
	−1	+1	——					
To	**0**			——				
	1				——			
	2					——		
	3						——	
	4							——

Functional Maths Activity (continued)

Task 2

The manager works out that the average time for a lift to move from one floor to the next is actually 15 seconds when going down and 20 seconds when going up.

Complete this table to show the times for the lift to move between floors.

		From						
		−2	**−1**	**0**	**1**	**2**	**3**	**4**
	−2	———	15 sec					
	−1	20 sec	———					
To	**0**			———				
	1				———			
	2					———		
	3						———	
	4							———

Task 3

Work out the longest time that it would take someone to get from their room to reception using the lift.

Show clearly how you work out your answer.

HINTS AND TIPS

The answer is not 60 seconds.

4 Number: Percentages and ratio

4.1 Equivalent percentages, fractions and decimals

HOMEWORK 4A

Example 1   As a fraction $32\% = \frac{32}{100}$ which can be cancelled down to $\frac{8}{25}$

Example 2   As a decimal $65\% = 65 \div 100 = 0.65$

1 Write each percentage as a fraction in its lowest terms.
  a 10%   **b** 40%   **c** 25%   **d** 15%   **e** 75%   **f** 35%
  g 12%   **h** 28%   **i** 56%   **j** 18%   **k** 42%   **l** 6%

2 Write each percentage as a decimal.
  a 87%   **b** 25%   **c** 33%   **d** 5%   **e** 1%   **f** 72%
  g 58%   **h** 17.5%   **i** 8.5%   **j** 68.2%   **k** 150%   **l** 132%

3 Copy and complete the table.

Percentage	Fraction	Decimal
10%		
20%		
30%		
		0.4
		0.5
		0.6
	$\frac{7}{10}$	
	$\frac{8}{10}$	
	$\frac{9}{10}$	

4 If 45% of pupils walk to school, what percentage do not walk to school?

5 If 84% of the families in a village own at least one car, what percentage of the families do not own a car?

6 In a local election, of all the people who voted, 48% voted for Mrs Slater, 29% voted for Mr Rhodes and the remainder voted for Mr Mulley. What percentage voted for Mr Mulley?

FM 7 From his gross salary, Mr Hardy pays 20% Income Tax, 6% Superannuation and 5% National Insurance. What percentage is his net pay?

FM Functional Maths **AU** (AO2) Assessing Understanding **PS** (AO3) Problem Solving

PS 8 Approximately what percentage of each can is filled with oil?

a 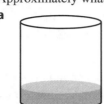 **b**  **c**

9 Write each fraction as a percentage.
  a $\frac{3}{4}$ **b** $\frac{2}{5}$ **c** $\frac{7}{20}$ **d** $\frac{3}{25}$ **e** $\frac{43}{50}$ **f** $\frac{3}{8}$

10 Write each decimal as a percentage.
  a 0.23   **b** 0.87   **c** 0.09   **d** 0.235   **e** 1.8   **f** 2.34

AU 11 Tom scored 68 marks out of a possible 80 marks in a geography test.
  a Write his score as a fraction in its simplest form.
  b Write his score as a decimal.
  c Write his score as a percentage.
  d His next test is of the same standard but is marked out of 50.
    How many marks out of 50 does he need to improve his percentage score?

4.2 Calculating a percentage of a quantity

HOMEWORK 4B

Example   Calculate 12% of 54 kg.

Method 1 $12 \div 100 \times 54 = 6.48$ kg
Method 2 Using a multiplier: $0.12 \times 54 = 6.48$ kg

1 What multiplier is equivalent to a percentage of:
  a 23%   **b** 70%   **c** 4%   **d** 120%?

2 What percentage is equivalent to a multiplier of:
  a 0.38   **b** 0.8   **c** 0.07   **d** 1.5?

3 Calculate the following.
  a 25% of £200   **b** 10% of £120   **c** 53% of 400 kg   **d** 75% of 84 cm
  e 22% of £84   **f** 71% of 250 g   **g** 24% of £3   **h** 95% of 320 m
  i 6% of £42   **j** 17.5% of £56   **k** 8.5% of 160 l   **l** 37.2% of £800

4 During one week at a Test Centre, 320 people took their driving test and 65% passed.
  How many people passed?

PS 5 A school has 250 pupils in each year and the attendance record on one day for each year
  group is shown below.
  Year 7 96%, Year 8 92%, Year 9 84%,   Year 10 88%, Year 11 80%
  The school has a target of 90% attendance overall. Did the school meet its target?

6 A certain type of stainless steel consists of 84% iron, 14% chromium and 2% carbon (by
  weight). How much of each is in 450 tonnes of stainless steel?

FM 7 VAT (Value Added Tax) is a tax that the Government adds to the price of goods sold. At the moment it is 17.5%. How much VAT will be added on to the following bills:

a a restaurant bill for £40   **b** a telephone bill for £82

c a car repair bill for £240?

FM 8 An insurance firm sells house insurance and the annual premiums are usually at a cost of 0.5% of the value of the house. What will be the annual premium for a house valued at £120 000?

9 A shop has a sale and reduces all prices by 10%. One week later the shop reduces prices by a further 10%. Have the prices reduced by 20% altogether? Show how you decide.

4.3 Ratio

HOMEWORK 4C

Example 1   Simplify 5 : 20.

5 : 20 = 1 : 4   (Divide both sides of the ratio by 5.)

Example 2   Simplify 20p : £2.

(Change to a common unit) 20p : 200p = 1 : 10

Example 3   A garden is divided into lawn and shrubs in the ratio 3 : 2.

The lawn covers $\frac{3}{5}$ of the garden and the shrubs cover $\frac{2}{5}$ of the garden.

1 Express each of the following ratios in their simplest form.

a 3 : 9   **b** 5 : 25   **c** 4 : 24   **d** 10 : 30   **e** 6 : 9

f 12 : 20   **g** 25 : 40   **h** 30 : 4   **i** 14 : 35   **j** 125 : 50

2 Express each of the following ratios of quantities in their simplest form. (Remember to change to a common unit where necessary.)

a £2 to £8   **b** £12 to £16   **c** 25 g to 200 g

d 6 miles : 15 miles   **e** 20 cm : 50 cm   **f** 80p : £1.50

g 1 kg : 300g   **h** 40 seconds : 2 minutes   **i** 9 hours : 1 day

j 4 mm : 2 cm

3 £20 is shared out between Bob and Kathryn in the ratio 1 : 3.

a What fraction of the £20 does Bob receive?

b What fraction of the £20 does Kathryn receive?

4 In a class of students, the ratio of boys to girls is 2 : 3.

a What fraction of the class is boys?

b What fraction of the class is girls?

PS FM 5 Pewter is an alloy containing lead and tin in the ratio 1 : 9.

a What fraction of pewter is lead?

b What fraction of pewter is tin?

c A foundry has 30 tonnes of lead and 90 tonnes of tin. How much pewter can it make?

AU 6 Roy wins two-thirds of his snooker matches. He loses the rest.
What is his ratio of wins to losses?

PS 7 In the 2009 Ashes cricket series, the numbers of wickets taken by Steve Harmison and Monty Panesar were in the ratio 5 : 1.
The ratio of the number of wickets taken by Graham Onions to those taken by Steve Harmison was 2 : 1.
What fraction of the wickets taken by these three bowlers was by Monty Panesar?

HOMEWORK 4D

Example   Divide £40 between Peter and Hitan in the ratio 2 : 3.

Changing the ratio to fractions gives
Peter's share = $\frac{2}{5}$ and Hitan's share = $\frac{3}{5}$
So, Peter receives $\frac{2}{5} \times £40 = £16$ and Hitan receives $\frac{3}{5} \times £40 = £24$.

1 Divide each of the following amounts in the given ratios.
  a £10 in the ratio 1 : 4   **b** £12 in the ratio 1 : 2
  c £40 in the ratio 1 : 3   **d** 60 g in the ratio 1 : 5
  e 10 hours in the ratio 1 : 9

2 The ratio of female to male members of a sports centre is 3 : 1. The total number of members of the centre is 400.
  a How many members are female?   **b** How many members are male?

3 A 20 metre length of cloth is cut into two pieces in the ratio 1 : 9. How long is each piece?

4 Divide each of the following amounts in the given ratios.
  a 25 kg in the ratio 2 : 3   **b** 30 days in the ratio 3 : 2
  c 70 m in the ratio 3 : 4   **d** £5 in the ratio 3 : 7
  e 1 day in the ratio 5 : 3

5 James collects beer mats and the ratio of British mats to foreign mats is 5 : 2. He has 1400 beer mats in his collection. How many foreign beer mats does he have?

6 Patrick and Jane share out a box of sweets in the ratio of their ages. Patrick is 9 years old and Jane is 11 years old. If there are 100 sweets in the box, how many does Patrick get?

AU 7 For her birthday Reena is given £30. She decides to spend four times as much as she saves. How much does she save?

FM 8 Mrs Megson calculates that her quarterly electric and gas bills are in the ratio 5 : 6. The total she pays for both bills is £66. She has saved £33 of gas stamps and £33 of electricity stamps to pay the bill. Will she need to pay any extra and if so, how much?

HINTS AND TIPS

Gas stamps cannot be used to pay for electricity.

9 You can simplify a ratio by changing it into the form 1 : *n*. For example, 5 : 7 can be rewritten as 5 : 7 = 1 : 1.4 by dividing each side of the ratio by 5. Rewrite each of the following ratios in the form 1 : *n*.
  a 2 : 3   **b** 2 : 5   **c** 4 : 5   **d** 5 : 8   **e** 10 : 21

PS 10 The amount of petrol and diesel sold at a garage is in the ratio 2 : 1. One- tenth of the diesel sold is bio-diesel.

What fraction of all the fuel sold is bio-diesel?

HOMEWORK 4E

> **Example** Two business partners, John and Ben, divided their total profit in the ratio 3 : 5. John received £2100. How much did Ben get?
>
> John's £2100 was $\frac{3}{8}$ of the total profit.
> So, $\frac{1}{8}$ of the total profit = £2100 ÷ 3 = £700.
> Therefore, Ben's share, which was $\frac{5}{8}$, amounted to £700 × 5 = £3500.

1 Peter and Margaret's ages are in the ratio 4 : 5. If Peter is 16 years old, how old is Margaret?

2 Cans of lemonade and packets of crisps were bought for the school disco in the ratio 3 : 2. The organiser bought 120 cans of lemonade. How many packets of crisps did she buy?

FM 3 In his restaurant, Manuel is making 'Sangria', a drink made from red wine and iced soda water, mixed in the ratio 2 : 3. Manuel uses 10 litres of red wine.
   a How many litres of soda water does he use?
   b How many litres of Sangria does he make?

4 Cupro-nickel coins are minted by mixing copper and nickel in the ratio 4 : 1.
   a How much copper is needed to mix with 20 kg of nickel?
   b How much nickel is needed to mix with 20 kg of copper?

5 The ratio of male to female spectators at a school inter-form football match is 2 : 1. If 60 boys watched the game, how many spectators were there in total?

6 Marmalade is made from sugar and oranges in the ratio 3 : 5. A jar of 'Savilles' marmalade contains 120 g of sugar.
   a How many grams of oranges are in the jar?
   b How many grams of marmalade are in the jar?

AU 7 Each year Abbey School holds a sponsored walk for charity. The money raised is shared between a local charity and a national charity in the ratio 1 : 2. Last year the school gave £2000 to the local charity.
   a How much did the school give to the national charity?
   b How much did the school raise in total?

PS 8 Fred's blackcurrant juice is made from four parts blackcurrant and one part water.
Jodie's blackcurrant juice is made from blackcurrant and water in the ratio 7 : 2.
Which juice contains the greater proportion of blackcurrant?
Show how you work out your answer.

4.4 Best buys

Example   There are two different-sized packets of Whito soap powder at a supermarket. The medium size contains 800 g and costs £1.60 and the large size contains 2.5 kg and costs £4.75. Which is the better buy?

Find the weight per unit cost for both packets.
Medium: 800 ÷ 160 = 5 g per pence
Large: 2500 ÷ 475 = 5.26 g per pence

From these we see that there is more weight per pence with the large size, which means that the large size is the better buy.

FM **1**   Compare the prices of the following pairs of products and state which of each pair is the better buy.

a

Mouthwash:
£1.50 for a bottle
£2.50 for a twin-pack

b

Deodorant:
£2.20 for 1
£4.45 for 2

c

Dusters:
49p for 5
95p for 10

d

Peas:
98p for 250 grams
£2.75 for 750 grams

 2   Compare the following pairs of product and state which is the better buy and why.

a   Tomato ketchup: a medium bottle which is 200 g for 55p or a large bottle which is 350 g for 87p.

b   Milk chocolate: a 125 g bar at 77p or a 200 g bar at 92p.

c   Coffee: a 750 g tin at £11.95 or a 500 g tin at £7.85.

d   Honey: a large jar which is 900 g for £2.35 or a small jar which is 225 g for 65p.

FM 3 Boxes of Wetherels teabags are sold in three different sizes.

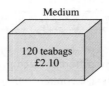 

Small
80 teabags
£1.44

Medium
120 teabags
£2.10

Large
200 teabags
£3.25

Which size of teabags gives the best value for money?

  4 Bottles of Cola are sold in different sizes. Copy and complete the table.

Size of bottle	Price	Cost per litre
$\frac{1}{2}$ litre	36p	
$1\frac{1}{2}$ litres	99p	
2 litres	£1.40	
3 litres	£1.95	

Which bottle gives the best value for money?

 AU 5 The following special offers are being promoted by a supermarket.

Only
£1.99
each
Cornflakes
750 g
£1.99

Cornflakes
500 g
£1.69
Buy 3
for the
price of 2

Which offer is the better value for money? Explain why.

   PS 6 Hannah scored 17 out of 20 in a test. John scored 40 out of 50 on a test of the same standard.
Who got the better mark?

4.5 Speed, time and distance

HOMEWORK 4G

The relationship between speed, time and distance can be expressed in three ways.

$$\text{Distance} = \text{Speed} \times \text{Time} \qquad \text{Speed} = \frac{\text{Distance}}{\text{Time}} \qquad \text{Time} = \frac{\text{Distance}}{\text{Speed}}$$

Example Sean is going to drive from Newcastle upon Tyne to Nottingham, a distance of 190 miles. He estimates that he will drive at an average speed of 50 mph. How long will it take him?

Sean's time = $\frac{190}{50}$ = 3.8 hours
Change the 0.8 hours to minutes by multiplying by 60, to give 48 minutes.
So, the time for Sean's journey will be 3 hours 48 minutes.

Remember When you calculate a time and get a decimal answer, do not mistake the decimal part for minutes. You must either:

● leave the time as a decimal number and give the unit as hours, or
● change the decimal part to minutes by multiplying it by 60
 (1 hour = 60 minutes) and give the answer in hours and minutes.

1 A cyclist travels a distance of 60 miles in 4 hours. What was her average speed?

2 How far along a motorway will you travel if you drive at an average speed of 60 mph for 3 hours?

3 Mr Baylis drives on a business trip from Manchester to London in $4\frac{1}{2}$ hours. The distance he travels is 207 miles. What is his average speed?

FM 4 The distance from Leeds to Birmingham is 125 miles. The train I catch travels at an average speed of 50 mph. If I catch the 11.30am train in Leeds, at what time would I expect to be in Birmingham?

5 Copy and complete the following table.

	Distance travelled	Time taken	Average speed
a	240 miles	8 hours	
b	150 km	3 hours	
c		4 hours	5 mph
d		$2\frac{1}{2}$ hours	20 km/h
e	1300 miles		400 mph
f	90 km		25 km/h

FM 6 A coach travels at an average speed of 60 km/h for 2 hours on a motorway and then slows down in a town centre to do the last 30 minutes of a journey at an average speed of 20 km/h.
a What is the total distance of this journey?
b What is the average speed of the coach over the whole journey?

FM AU 7 Hilary cycles to work each day. She cycles the first 5 miles at an average speed of 15 mph and then cycles the last mile in 10 minutes.
a How long does it take Hilary to get to work?
b What is her average speed for the whole journey?

FM 8 Martha drives home from work in 1 hour 15 minutes. She drives home at an average speed of 36 mph.
a Change 1 hour 15 minutes to decimal time in hours.
b How far is it from Martha's work to her home?

PS AU 9 A tram route takes 15 minutes at an average speed of 16 mph.
The same journey by car is 2 miles longer.
How fast would a car need to travel to arrive at the destination in the same amount of time?

Functional Maths Activity

The cost of going to work

Miss Jones

- Miss Jones is 23 years old and lives in Bramley.
- She works 20 miles from home, in Aston.
- She is a manager in a small company and earns £18 000 per year.
- She works from Monday to Friday each week.
- She has four weeks holiday per year.
- She always takes two of these holiday weeks in July every year.
- She travels to work by train each day using a monthly ticket.
- She has a 16–25 Railcard.
- In July she buys weekly tickets.
- The journey takes 45 minutes each way.
- She uses the local sandwich shop for lunch each day.

Rail Fares
Bramley to Aston
7-Day ticket £56.40
Monthly ticket £217.35
3-month ticket £652.05

16–25 Railcard — £26 for a whole year

Aged 16–25 or a full-time student aged 26 or over?

Save 1/3 on most rail fares throughout Great Britain

16–25 Railcard discounts now apply to *all*
Standard and First Class Advance fares.

Sandwich shop
Small sandwiches ~ £2.50
Large Sandwiches ~ £3.30
Pay weekly for your sandwiches
and get Friday free!

Mr Smith

- Mr Smith is 45 years old and lives in Sunnyside.
- He works 10 miles from home, in Todwick.
- He is a maintenance worker in the same small company and earns £12000 per year.
- He works from Monday to Saturday each week.
- He has six weeks holiday per year.
- He travels to work by bus each day, using a daily return ticket.
- The journey takes 30 minutes each way.
- He has lunch in the work canteen each day.
- He always has the set meal, plus a drink and two portions of extra vegetables.

Bus fares
Todwick to Aston
Single £2.35
Return £3.20

WORKS CANTEEN

SET MEAL --- £4
DRINKS   --- 75P EACH
EXTRA VEGETABLES 50P PER PORTION

Functional Maths Activity (continued)

Task 1

Answer the following questions about Miss Jones.

1 How many weeks does she work in a year?
2 How much is she paid each month?
3 How much does she pay for a monthly rail ticket?
4 Why does she buy weekly tickets in July?
5 How much does she spend on travel to work, including the cost of the Railcard, in one year?
6 How much would she pay at the sandwich shop if she pays weekly for small sandwiches?
7 What proportion of the cost of the sandwiches is she saving?
8 How much more per week would it cost her if she had large sandwiches?
9 What is the ratio of the cost of small sandwiches to large sandwiches? Give your answer in its simplest form.
10 One-third of her salary is spent in taxes. How much does she have left after tax?

Task 2

Use your answers to **Task 1** to help you to work out how much money Miss Jones has left after deducting taxes, travelling and meal costs from her salary.
Give your answer as a monthly amount.

Task 3

Work the time that Miss Jones spends travelling to and from work each year.

Task 4

Work out how much money Mr Smith has left after deducting taxes, travelling and meal costs from his salary.

Task 5

- Imagine that you live 15 miles from work.
- Decide costs for travelling to work by train or bus.
- Decide what you will eat at lunchtime.
- Decide what your salary will be.
- If the salary is low, deduct one-third of the salary for taxes.
- If the salary is high, the ratio of tax to remaining pay is 2 : 1.
- Work out how much money you will have left after deducting taxes, travelling and meal costs from your salary.

5 Number: Further number skills

5.1 Long multiplication

1 24×13 **2** 33×17 **3** 54×42 **4** 89×23 **5** 58×53

6 176×14 **7** 235×16 **8** 439×21 **9** 572×35 **10** 678×57

FM 11 Andy needs enough tiles to cover 12 m² in his bathroom. It takes 25 tiles to cover 1 m². It is recommended to buy 20% more to allow for cutting.
Tiles are sold in boxes of 16. Andy buys 24 packs. Does he have enough tiles?

FM 12 A TV rental shop buys TVs for £110 each.
The shop need to make at least 10% on each TV to cover overheads.
On average each TV is rented for 40 weeks at £3.50 per week.
Does the shop cover its overheads?

FM 13 Mrs Woodhead saves £14 per week towards her fuel bills.
Her annual fuel bills are £900.
Does she save enough?

FM 14 Sylvia has a part-time job and is paid £18 for every day she works. Last year she worked for 148 days. She saves all her income towards a summer holiday that costs £2000. She changes the extra into Euros at an exchange rate of £1 = €1.20. How many Euros does she get?

PS 15 A concert hall has 48 rows of seats with 32 seats in a row. What is the maximum capacity of the hall?

PS AU 16 A room measuring 6 metres by 8 metres is to be carpeted. The carpet costs £19 per square metre.
a Estimate the cost of the carpet.
b Calculate the exact cost of the carpet.

5.2 Long division

1 $312 \div 13$ **2** $480 \div 15$ **3** $697 \div 17$ **4** $792 \div 22$ **5** $806 \div 26$

6 $532 \div 28$ **7** $736 \div 32$ **8** $595 \div 35$ **9** $948 \div 41$ **10** $950 \div 53$

FM 11 The organiser of a church fete needs 1000 balloons. She has a budget of £30. Each packet contains 25 balloons and costs 85p. Does she have enough?

FM 12 The annual subscription fee to join a Fishing Club is £42. The treasurer of the club has collected £1134 in fees. The annual rental for the fishing pond is £2000. How many more members need to join so that they can pay the rental?

FM **13** A coach firm charges £504 for 36 people to go Christmas shopping on a day trip to Calais. Mary takes £150. The cost of the coach is shared equally between the passengers. Mary wants to buy a game costing €50 for each of her 4 grandchildren. The exchange rate is £1 = €1.25. Does she have enough?

PS AU **14** Allan is a market gardener and has 420 bulbs to plant. He plants them out in rows with 18 bulbs to a row. How many complete rows will there be?

5.3 Arithmetic with decimal numbers

HOMEWORK 5C

Example 1	5.852 will round off to 5.85 to two decimal places
Example 2	7.156 will round off to 7.16 to two decimal places
Example 3	0.284 will round off to 0.3 to one decimal place
Example 4	15.3518 will round off to 15.4 to one decimal place

1 Round each of the following numbers to one decimal place.
a 3.73	**b** 8.69	**c** 5.34	**d** 18.75	**e** 0.423
f 26.288	**g** 3.755	**h** 10.056	**i** 11.08	**j** 12.041

2 Round each of the following numbers to two decimal places.
a 6.721	**b** 4.457	**c** 1.972	**d** 3.485	**e** 5.807
f 2.564	**g** 21.799	**h** 12.985	**i** 2.302	**j** 5.555

3 Round each of the following to the number of decimal places indicated.
a 4.572 (1 dp)	**b** 0.085 (2 dp)	**c** 5.7159 (3 dp)	**d** 4.558 (2 dp)
e 2.099 (2 dp)	**f** 0.7629 (3 dp)	**g** 7.124 (1 dp)	**h** 8.903 (2 dp)
i 23.7809 (3 dp)	**j** 0.99 (1 dp)		

4 Round each of the following to the nearest whole number.
a 6.7	**b** 9.3	**c** 2.8	**d** 7.5	**e** 8.38
f 2.82	**g** 2.18	**h** 1.55	**i** 5.252	**j** 3.999

FM **5** Trevor buys the following car accessories:
Shampoo £4.99; Wax £7.29; Wheel cleaner £4.81; Dusters £1.08.
By rounding each price to the nearest pound, work out an estimate of the total cost of the items.

AU **6** Which of the following are correct roundings of the number 9.281?
9   9.2   9.28   9.3   9.30

PS **7** When a number is rounded to two decimal places, the answer is 6.14.
Which of these could be the number?
6.140   6.143   6.148   6.15

HOMEWORK 5D

Example

Work out $4.2 + 8 + 12.93$. Set out the sum as follows:

Remember to keep the points in the same column.

$$\begin{array}{r} 4.20 \\ 8.00 \\ + 12.93 \\ \hline 25.13 \\ \hline \small{1\ 1} \end{array}$$

1 Work out each of these.

    a $7.3 + 2.6$ **b** $15.7 + 5.6$ **c** $33.5 + 6.8$ **d** $8.5 + 4.82$

    e $3.26 + 4.5$ **f** $2.75 + 9.84$ **g** $24.5 + 6.3$ **h** $8.4 + 12.8$

    i $13.75 + 8.5$ **j** $7.08 + 0.7$ **k** $7 + 2.96 + 3.1$ **l** $8.5 + 7.36 + 12.1$

2 Work out each of these.

    a $5.8 - 3.4$ **b** $7.3 - 2.8$ **c** $4.6 - 2.7$ **d** $9.7 - 4.7$

    e $8.35 - 4.24$ **f** $9.74 - 3.81$ **g** $9.04 - 5.72$ **h** $3.62 - 1.85$

    i $6 - 3.3$ **j** $8 - 7.4$ **k** $12 - 3.2$ **l** $7.2 - 4.72$

AU 3 Olivia has £3.80 left after buying a toy for £8.99.

    a   How much did she have before she bought the toy?

    b   She then goes home by tram.

    After paying her tram fare she has £1.05 left.

    How much is the tram fare?

FM 4 A piece of cloth is 2.7 metres long.

    Is it possible to get two pieces of lengths 1.8 metres and 0.8 metres from the cloth?

PS 5 Copy and complete the following.

    a $16.3 + \ldots\ldots = 20.7$

    b $41.5 + \ldots\ldots = 67.9$

    c $\ldots\ldots + 3.4 = 9.2$

    d $\ldots\ldots + 7.35 = 8.26$

6 Copy and complete the following.

    a $21.3 - \ldots\ldots = 13.8$

    b $70.5 - \ldots\ldots = 48.2$

    c $\ldots\ldots - 6.3 = 8.4$

    d $\ldots\ldots - 12.5 = 8.7$

HOMEWORK 5E

Example 1 $4.5 \times 3 =$

$$\begin{array}{r} 4.5 \\ \times\ \ 3 \\ \hline 13.5 \\ \hline \small{1} \end{array}$$

Example 2 $8.25 \div 5 =$

$$\begin{array}{r} 1.6\ 5 \\ 5\ \overline{\smash{\big)}\ 8.^32^25} \end{array}$$

Example 3 $5.7 \div 2 =$

$$\begin{array}{r} 2.8\ 5 \\ 2\ \overline{\smash{\big)}\ 5.^17^10} \end{array}$$

1 Evaluate each of these.

    a 2.3×3 **b** 4.8×2 **c** 4.6×4 **d** 15.3×5 **e** 26.4×8

2 Evaluate each of these.

    a 2.14×2 **b** 3.45×3 **c** 5.47×6 **d** 4.44×8 **e** 0.25×9

3 Evaluate each of these.

    a $4.8 \div 2$ **b** $7.6 \div 4$ **c** $7.2 \div 3$ **d** $7.35 \div 5$ **e** $0.78 \div 6$

4 Evaluate each of these.

    a $4.5 \div 2$ **b** $7.2 \div 5$ **c** $3.4 \div 4$ **d** $13.1 \div 5$ **e** $6.3 \div 8$

FM 5 Crisps are sold in packs of six for £1.32 or packs of eight for £1.92. Which are better value?

FM 6 Steve takes his wife and three children on a day trip by train to London.

    a   The tickets cost £26.60 for each adult and £12.85 for each child. How much do the tickets cost Steve altogether?

    b   While they are in London, they decide to go on a bus tour.

           Tickets are £22 for adults and £10 for children.

           Steve only has enough money if he can negotiate £2 off the price of each ticket.

           How much money does he have?

HOMEWORK 5F

Example   Evaluate $4.27 \times 34 =$ 4.27

                               $\times$ 34

                           17.08   (multiply by 4)

                        128.10   (multiply by 3 and keep points in same column)

                        145.18

1 Evaluate each of these.

    a 3.12×14 **b** 5.24×15 **c** 1.36×22 **d** 7.53×25 **e** 27.1×32

2 Find the total cost of each of the following purchases.

    a   24 litres of petrol at £1.22 per litre

    b   18 pints of milk at £0.56 per pint.

    c   14 magazines at £2.75 per copy.

FM 3 The table shows the exchange rate for various currencies

Currency	Exchange rate
Euro (€)	£1 = €1.15
American dollar ($)	£1 = $1.55
Swiss franc (F)	£1 = 1.65F

    a   Douglas changes £25 into Euros. How many Euros does he get?

    b   Martin changes £32 into dollars. How many dollars does he get?

    c   Pauline changes £45 into francs. How many francs does she get?

PS 4 Jordan says that he can change multiplications into easier multiplications by changing the multiplication around as shown:

To work out 0.8×72, he does 7.2×8

Use his method to work out 0.6×21

AU 5 Pola is paid £9.71 per hour for working 30 hours each week.

She saves half of her pay.

How many weeks will it take her to save £1000?

HOMEWORK 5G

To multiply one decimal number by another decimal number:
- First, do the whole calculation as if the decimal points were not there.
- Then, count the total number of decimal places in the two decimal numbers. This gives the number of decimal places in the answer.

Example   Evaluate 3.42×0.2

Ignoring the decimal points gives the following calculation: $342 \times 2 = 684$

Now, 3.42 has two decimal places and 0.2 has one decimal place. So, the total number of decimal places in the answer is three, which gives $3.42 \times 0.2 = 0.684$

1 Evaluate each of these.

a	2.3×0.2	**b**	5.2×0.3	**c**	4.6×0.4	**d**	0.2×0.3	**e**	0.4×0.7
f	0.5×0.5	**g**	12.6×0.6	**h**	7.2×0.7	**i**	1.4×1.2	**j**	2.6×1.5

FM 2 Kerry wants to buy a plank of wood 6.5 metres long. It costs £1.25 per metre. She has £10. Can she afford it?

3 For each of the following:
i   estimate the answer by first rounding off each number to the nearest whole number.
ii   calculate the exact answer, and then, by doing a subtraction, calculate how much out your answer to part **i** is.

a	3.7×2.4	**b**	4.8×3.1	**c**	5.1×4.2	**d**	6.5×2.5

PS AU 4 **a** Use any method to work out 15×16
b Use your answer to work out:
   i 1.5×1.6
   ii 0.75×3.2
   iii 4.5×1.6

5.4 Multiplying and dividing with negative numbers

HOMEWORK 5H

1 Write down the answers to the following.

a	-2×4	**b**	-3×6	**c**	-5×7	**d**	-3×-4	**e**	-8×-2
f	$-14 \div -2$	**g**	$-16 \div -4$	**h**	$25 \div -5$	**i**	$-16 \div -8$	**j**	$-8 \div -4$
k	3×-7	**l**	6×-3	**m**	7×-4	**n**	-3×-9	**o**	-7×-2
p	$28 \div -4$	**q**	$12 \div -3$	**r**	$-40 \div 8$	**s**	$-15 \div -3$	**t**	$50 \div -2$
u	-3×-8	**v**	$42 \div -6$	**w**	7×-9	**x**	$-24 \div -4$	**y**	-7×8

2 Write down the answers to the following.

a	$-2 + 4$	**b**	$-3 + 6$	**c**	$-5 + 7$	**d**	$-3 + -4$	**e**	$-8 + -2$
f	$-14 - -2$	**g**	$-16 - -4$	**h**	$25 - -5$	**i**	$-16 - -8$	**j**	$-8 - -4$
k	$3 + -7$	**l**	$6 + -3$	**m**	$7 + -4$	**n**	$-3 + -9$	**o**	$-7 + -2$
p	$28 - -4$	**q**	$12 - -3$	**r**	$-40 - 8$	**s**	$-15 - -3$	**t**	$50 - -2$
u	$-3 + -8$	**v**	$42 - -6$	**w**	$7 + -9$	**x**	$-24 - -4$	**y**	$-7 + 8$

3 What number do you multiply -5 by to get the following?

a	25	**b**	-30	**c**	50	**d**	-100	**e**	75

AU 4   **a**   Work out 7 × –3.
      b   The average temperature drops by 3°C every day for a week. How much has the temperature dropped altogether by the end of the week?
      c   The temperature drops by 5°C for the next two days. Write down the calculation to work the total drop in temperature over these two days.

PS 5   Put these calculations in order from lowest to highest.
        –20 ÷ 2   –8 × –1   –24 ÷ –6   –7 × 2

5.5 Approximation of calculations

HOMEWORK 5I

1   Round each of the following numbers to one significant figure.

a	46 313	**b**	57 123	**c**	30 569	**d**	94 558	**e**	85 299
f	54.26	**g**	85.18	**h**	27.09	**i**	96.432	**j**	167.77
k	0.5388	**l**	0.2823	**m**	0.005 84	**n**	0.047 85	**o**	0.000 876
p	9.9	**q**	89.5	**r**	90.78	**s**	199	**t**	999.99

AU 2   What is the least and the greatest number of people that can be found in these towns?
      Hellaby   population 900 (to one significant figure)
      Hook   population 650 (to two significant figures)
      Hundleton   population 1050 (to three significant figures)

PS 3   A baker estimates that she has baked 100 loaves. She is correct to one significant figure.
      She sells two loaves and now has 90 loaves to one significant figure.
      How many could she have had to start with?
      Work out all possible answers.

AU 4   There are 500 cars in a car park to one significant figure.
      What is the least possible number of cars that could enter the car park so that there are 700 cars in the car park to one significant figure?

HOMEWORK 5J

1   Find approximate answers to the following sums.

a	4324 × 6.71	**b**	6170 × 7.311	**c**	72.35 × 3.142
d	4709 × 3.81	**e**	63.1 × 4.18 × 8.32	**f**	320 × 6.95 × 0.98
g	454 ÷ 89.3	**h**	26.8 ÷ 2.97	**i**	4964 ÷ 7.23
j	316 ÷ 3.87	**k**	2489 ÷ 48.58	**l**	63.94 ÷ 8.302

2   Find the approximate monthly pay of the following people with the following annual salaries.
      a   Joy £47 200   **b**   Amy £24 200   **c**   Tom £19 135

3   Find the approximate annual pay of these brothers who earn the following amounts.
      a   Trevor £570 a week   **b**   Brian £2728 a month

AU 4   A groundsman bought 350 kg of seed at a cost of £3.84 per kg.
      a   Find the approximate total cost of this seed.
      b   Did he make a profit?

FM 5   A greengrocer sells a box of 250 apples for £47. Approximately how much did each apple sell for?

6 Keith runs about 15 km every day. Approximately how far does he run in:
    a   a week   **b**   a month   **c**   a year?

7 A litre of creosote will cover an area of about 6.8 m^2. Approximately how many litre cans will I need to buy to creosote a fence with a total surface area of 43 m^2?

PS 8 A tour of London sets off at 1013 and costs £21. It returns at 1208.
Approximately how much does the tour cost per hour?

HOMEWORK 5K

FM 1 Round each of the following to give sensible answers.
    a   Kris is 1.6248 metres tall.
    b   It took me 17 minutes 48.78 seconds to cook the dinner.
    c   My rabbit weighs 2.867 kg.
    d   The temperature at the bottom of the ocean is 1.239 °C.
    e   There were 23 736 people at the game yesterday.

2 How many jars each holding 119 cm^3 of water can be filled from a 3-litre flask?

AU 3 If I walk at an average speed of 62 metres per minute, how long will it take me to walk a distance of 4 km?

FM 4 Helen earns £31 500 a year. She works five days a week for 45 days a year. How much does she earn a day?

5 10 grams of Gold cost £2.17. How much will one kilogram of Gold cost?

FM 6 Rewrite the following article using sensible numbers:
I left home at eleven and a half minutes past two, and walked for 49 minutes.
The temperature was 12.7623 °C. I could see an aeroplane overhead at 2937.1 feet.
Altogether I had walked 3.126 miles.

Functional Maths Activity

Shopping for the elderly

Thelma is a warden for the elderly.
She collects shopping for those who need help.

White Bread
(800g)
£1.21

White Bread
(400g)
79p

Bread Rolls
79p
for 6

Brown Bread
(400g)
83p

Brown Bread
(800g)
£1.45

Bananas
57p per
kilogram

Cabbage
68p each

Carrots
68p per
kilogram

Apples
£1.97 per
kilogram

Oranges
27p each

Potatoes
Small Pack
£1.35

Potatoes
Large Pack
£2.95

Sausages
£3.60
for 10 (450g)

Bacon
£1.80 for
6 slices (240g)

Beef joints
from
£4.00

Eggs
Large Free Range
32p for 6

Harriet, Doris and Wilf are three of the elderly people.

Harriet and Wilf have shopping lists.

Doris is a vegetarian and does not eat meat. She does not have a shopping list.

She gives Thelma £20 and asks her to buy what she can.

Harriet

Small white loaf
A beef joint (only spend about £5 on this)
6 eggs
2 oranges
Small pack potatoes

Wilf

Large pack potatoes
Large brown bread
6 bread rolls
About 5 sausages
½ kg bananas
½ kg apples

Tasks

Work out the costs of Harriet's and Wilf's shopping.
Try to buy a sensible range of food for Doris.
Do not forget that she does not eat meat.
Do not forget you must not spend more than £20
Show clearly any calculations that you make or any checking that you do.

6 Measures: Systems of measurement

6.1 Reading scales

1 Read the values from the following scales. Remember to state the units.

a

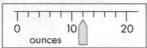

b

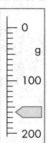

c

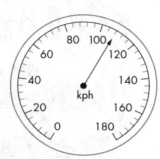

d

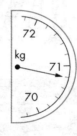

2 Read the temperatures shown on each of these thermometers.

a  **b** 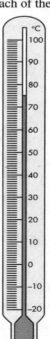 **c**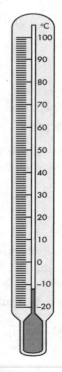

3   Read the values shown on these scales.

a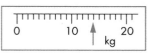

b

c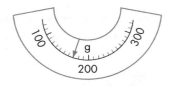

d

4   The speedometer below shows speeds in mph and km/h.

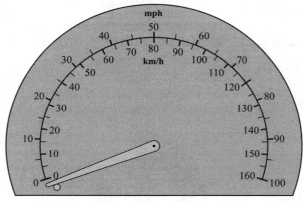

a   Use the scale on the speedometer to change 50mph into km/h.

b   Use the scale on the speedometer to change 120km/h into mph.

FM 5   Joe is going on holiday. He is allowed to take a suitcase weighing 15 kg.

Joe stands on the bathroom scales first on his own and then holding his suitcase.

Is his suitcase too heavy?

PS 6   Each of these three lines is 10 cm long.

Copy the lines and mark each one with an arrow pointing to a value of 30.

AU 7 The diagram shows part of a scale.

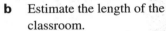

Lisa says that the arrow is pointing to 0.23 kg.
Explain why she is wrong.

6.2 Sensible estimates

HOMEWORK 6B

1 The picture shows a teacher standing by a whiteboard in a classroom.

    a Estimate the height of the door.
    b Estimate the length of the classroom.
    c Estimate the length and width of the whiteboard.

AU 2 Estimate the height of the giraffe.

3

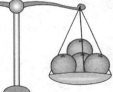

    a Estimate the weight of one apple.
    b Estimate the weight of one orange.

PS **4** The height of The London Eye is 135 m.
Use this information to estimate the height of:

a The Eiffel Tower

b Sears Tower

c Warszawa Radio Mast.

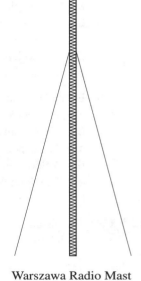

London Eye   Eiffel Tower   Sears Tower   Warszawa Radio Mast

FM **5** A builder wants to estimate the height of a lighthouse.

Use the picture to estimate the height.

6.3 Systems of measurement

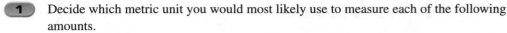
HOMEWORK 6C

1 Decide which metric unit you would most likely use to measure each of the following amounts.
  a The height of your best friend.
  b The distance from school to your home.
  c The thickness of a CD.
  d The weight of your maths teacher.
  e The amount of water in a lake.
  f The weight of a slice of bread.
  g The length of a double-decker bus.
  h The weight of a kitten.

2 Estimate the approximate metric length, weight or capacity of each of the following.
  a This book (both length and weight).
  b The length of the road you live on. (You do not need to walk along it all.)
  c The capacity of a bottle of wine (metric measure).
  d A door (length, width and weight).
  e The diameter of a £1 coin, and its weight.
  f The distance from your school to the Houses of Parliament (London).

AU 3 The distance from Sheffield to Tintagel is shown on a website as 294 miles.
Explain why this unit is used instead of metres.

FM 4 Sarah makes a living cleaning windows of houses. She has three sets of ladders which extend to 2 metres, 4 metres and 6 metres. Which one should she use to clean the upper windows of a two-storey house? Give a reason for your choice.

6.4 Metric units

HOMEWORK 6D

Length	10 mm = 1 cm, 1000 mm = 100 cm = 1 m, 1000 m = 1 km
Weight	1000 gm = 1 kg , 1000 kg = 1 t
Capacity	10 ml = 1 cl, 1000 ml = 100 cl = 1 litre
Volume	1000 litres = 1 m^3, 1 ml = 1 cl^3

1 Fill in the gaps using the information above.
  a 155 cm = ... m
  b 95 mm = ... cm
  c 780 mm = ... m
  d 3100 m = ... km
  e 310 cm = ... m
  f 3050 mm = ... m
  g 156 mm = ... cm
  h 2180 m = ... km
  i 1070 mm = ... m
  j 1324 cm = ... m
  k 175 m = ... km
  l 83 mm = ... m
  m 620 mm = ... cm
  n 2130 cm = ... m
  o 5120 m = ... km
  p 8150 g = ... kg
  q 2300 kg = ... t
  r 32 ml = ... cl
  s 1360 ml = ... l
  t 580 cl = ... l
  u 950 kg = ... t

2 Fill in the gaps using the information at the start of Homework 7B.
  a 120 g = ... kg
  b 150 ml = ... l
  c 350 cl = ... l
  d 540 ml = ... cl
  e 2060 kg = ... t
  f 7500 ml = ... l
  g 3800 g = ... kg
  h 605 cl = ... l
  i 15 ml = ... l
  j 6300 l = ... m^3
  k 45 ml = ... cm^3
  l 2350 l = ... m^3
  m 720 l = ... m^3
  n 8.2 m = ... cm
  o 71 km = ... m
  p 8.6 m = ... mm
  q 15.6 cm = ... mm
  r 0.83 m = ... cm
  s 5.15 km = ... m
  t 1.85 cm = ... mm
  u 2.75 m = ... cm

FM 3 Alesha wants to put up six shelves, each 65 cm long. She finds that she can buy planks of the correct width in three different lengths: 1200 mm, 1800 mm and 2400 mm. What should she buy to have the least amount of waste wood?

PS 4 Find how many square centimetres there are in a square kilometre.

AU 5 Could you pour all the water from a full 2 litre bottle into a container whose volume you know to be 101 cm^3?
Explain how you know.

6.5 Imperial units

HOMEWORK 6E

Length	12 inches = 1 foot, 3 feet = 1 yard, 1760 yards = 1 mile
Weight	16 ounces = 1 pound, 14 pounds = 1 stone, 2240 pounds = 1 ton
Capacity	8 pints = 1 gallon

1 Fill in the gaps using the information above.

a	5 feet = ... inches		**b**	5 yards = ... feet
c	3 miles = ... yards		**d**	6 pounds = ... ounces
e	5 stones = ... pounds		**f**	2 tons = ... pounds
g	4 gallons = ... pints		**h**	7 feet = ... inches
i	2 yards = ... inches		**j**	11 yards = ... feet
k	5 pounds = ... ounces		**l**	39 feet = ... yards
m	2 stones = ... ounces		**n**	4400 yards = ... miles
o	12 gallons = ... pints		**p**	2 miles = ... feet
q	84 inches = ... feet		**r**	48 ounces = ... pounds
s	21 feet = ... yards		**t**	22 400 pounds = ... tons
u	2 miles = ... inches		**v**	256 ounces = ... pounds
w	80 pints = ... gallons		**x**	280 pounds = ... stones
y	31 680 feet = ... miles		**z**	2 tons = ... ounces

PS 2 Find how many square feet are in a square mile.

FM 3 Running tracks in the UK and the USA used to be 400 yards long. How many times would you need to run around the track in a six-mile race?

AU 4 1 pound is approximately 450 grams.
Explain how you know that 1 tonne is lighter than 1 ton.

6.6 Conversion factors

HOMEWORK 6F

Length	Weight	Capacity
1 metre ≈ 39 inches	2.2 pounds ≈ 1 kilogram	1 litre ≈ 1.75 pints
1 foot ≈ 30 centimetres		1 gallon ≈ 4.5 litres
1 foot ≈ 12 inches		
5 miles ≈ 8 kilometres		

1 Change each of these weights in kilograms to pounds.
    a 6   **b** 8   **c** 15   **d** 32   **e** 45

2 Change each of these weights in pounds to kilograms. (Give each answer to 1 decimal place.)
    a 10   **b** 18   **c** 25   **d** 40   **e** 56

3 Change each of these capacities in litres to pints.
    a 2   **b** 8   **c** 25   **d** 60   **e** 75

4 Change each of these capacities in pints to litres. (Give each answer to the nearest litre.)
    a 7   **b** 20   **c** 35   **d** 42   **e** 100

5 Change each of these distances in miles to kilometres.
    a 20   **b** 30   **c** 50   **d** 65   **e** 120

6 Change each of these distances in kilometres to miles.
    a 16   **b** 24   **c** 40   **d** 72   **e** 300

7 Change each of these capacities in gallons to litres.
    a 5   **b** 12   **c** 27   **d** 50   **e** 72

8 Change each of these capacities in litres to gallons.
    a 18   **b** 45   **c** 72   **d** 270   **e** 900

9 Change each of these distances in metres to inches.
    a 2   **b** 5   **c** 8   **d** 10   **e** 12

10 Change each of these distances in feet to centimetres.
    a 3   **b** 5   **c** 7   **d** 10   **e** 30

11 Change each of these distances in inches to metres. (Give your answer to 1 decimal place.)
    a 48   **b** 52   **c** 60   **d** 75   **e** 100

FM 12 While on holiday in Iceland, I saw a road sign that said 'Blue Lagoon 26 km'.
I was in a coach travelling on a road with a speed limit of 40 km/h.
    a Approximately how many miles was I from the Blue Lagoon?
    b What was the speed limit in miles per hour?
    c If the coach travelled at this top speed, how long would it take us to get to the Blue Lagoon? Give your answer in minutes.

FM PS 13 While cycling up the Alps, Tom had a 60 km stretch to cycle in one day. He knew that he could only average 11.5 mph on these hills in the Alps.
How long would he expect this stretch to take him with no stops?

AU 14 How many cubic inches are in a five gallon drum?

Functional Maths Activity

Olde English Units

Most units of measurement were originally based on parts of the body.
In the old days in Britain, there were some very strange units used.

a   These units were used in the old days to measure cloth:

Cloth measures	
in	$2\frac{1}{4}$ inches = 1 nail
nl	4 nails = 1 quarter
qr	4 quarters = 1 yard (yd)
Fl. e	3 quarters = 1 Flemish ell
Eng. e	5 quarters = 1 English ell
Fr. e	6 quarters = 1 French ell
Sc. e	37 in = 1 Scotch ell

Work out the lengths of the following in centimetres.
Give your answers to the nearest centimetre.

i	A Flemish Ell	**ii**	An English Ell
iii	A French Ell	**iv**	A Scotch ell

b   The following units were used in the old days to measure heaped goods.

Heaped measures for lime, coals, fish, potatoes, fruit, etc.	
gall	2 gallons = 1 peck
pks	4 pecks = 1 bushel
bush	3 bushel = 1 sack
sks	12 sacks = 1 Chaldron (chal.)

c   These units were used in the old days to measure Yarn

Yarn Measures	
Cotton Yarn	Lint Yarn
54 inches = 1 thread	90 inches = 1 thread
80 threads = 1 skein or rap	120 threads = 1 cut
7 skeins = 1 hank	2 cuts = 1 heer
18 hanks = 1 spindle	6 heers = 1 hasp
	4 hasps = 1 spindle

How much longer, in metres, is a lint spindle than a cotton spindle?

Geometry and measures: Angles

7.1 Measuring and drawing angles

1 Use a protractor to find the size of each marked angle.

a

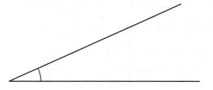

b

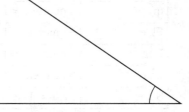

c

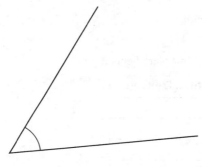

d

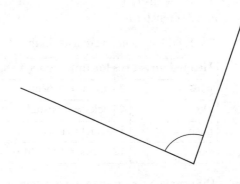

e

f

g

FM Functional Maths **AU** (AO2) Assessing Understanding **PS** (AO3) Problem Solving

h

2 Draw angles of the following sizes.

a 30° **b** 42° **c** 55° **d** 68°
e 75° **f** 140° **g** 164° **h** 245°

3 **a** Draw any three acute angles.
b Estimate their sizes. Record your results.
c Measure the angles. Record your results.
d Work out the difference between your estimate and your measurement for each angle.

4 Baden is using his compass and map to measure the angle between the North line and the summit of a mountain.

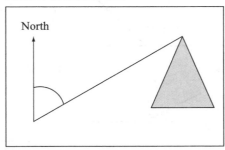

What is the size of the angle?

AU 5 Is angle ABC bigger than angle XYZ?

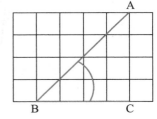 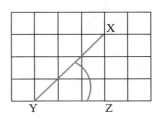

Give a reason for your answer.

AU 6 Which angle is the odd one out?
Give a reason for your answer.

a   b   c   d

PS 7 An acute angle is half the size of an obtuse angle.
Write down a possible size for the acute angle.

FM 8 When plumbers want to change the direction of a water pipe, they use a short section of
bent pipe called an elbow. The most common types are a 90 degree elbow or a 45 degree
elbow. Sketch what you think they look like.

FM 9 An acute angle is half the size of an obtuse angle.
Write down a possible size for the acute angle.

7.2 Angle facts

HOMEWORK 7B

> **Example**   Find the value of x in the diagram.
>
> These angles are around a point and add up to
> $360°$.
> So $x + x + 40 + 2x - 20 = 360°$
> $$4x + 20 = 360°$$
> $$4x = 340°$$
> $$x = 85°$$
>
>
> $x + 40$
> x
> $2x - 20$

1 Calculate the size of the angle marked x in each of these examples.

a   b   c

$120°$ x $135°$ x x $60°$

d   e   f

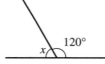

x $240°$ x $153°$ x $79°$

g   h   i

 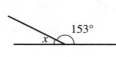 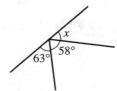

$30°$ x $50°$ x $30°$ x $58°$ $63°$

j

k

l

m

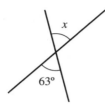

n

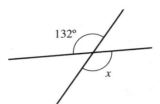

AU 2 Will these three angles fit together to make a straight line?

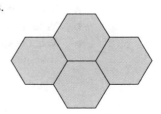

Explain your answer.

3 Calculate the value of x in each of these examples.

a

b

c

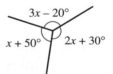

4 Calculate the value of x in each of these examples.

a

b

c

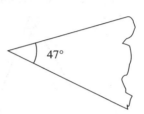

5 Calculate the value of x first and then find the size of angle y in each of these examples.

a

b

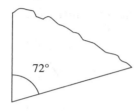

c

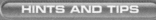

FM 6 These paving stones are in the shape of regular hexagons.
Explain why they fit together without leaving any gaps.

HINTS AND TIPS

All the angles inside a
regular hexagon are 120°.

PS **7** Angle BXC is three times the size of angle AXB and angle AXC is five times the size of angle AXB. Work out the size of the three angles.

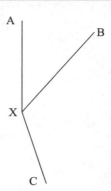

7.3 Angles in a triangle

HOMEWORK 7C

1 Find the size of the angle marked with a letter in each of these triangles.

a

b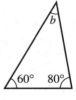

c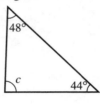

d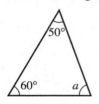

e

f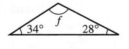

AU **2** State whether each of these sets of angles are the three angles of a triangle?
Explain your answers.
a 15°, 85° and 80° **b** 40°, 60° and 90° **c** 25°, 25° and 110°
d 40°, 40° and 100° **e** 32°, 37° and 111° **f** 61°, 59° and 70°

3 The three interior angles of a triangle are given in each case. Find the angle indicated by a letter.
a 40°, 70° and $a°$ **b** 60°, 60° and $b°$ **c** 80°, 90° and $c°$
d 65°, 72° and $d°$ **e** 130°, 45° and $e°$ **f** 112°, 27° and $f°$

4 In a triangle all the interior angles are the same.
a What size is each angle?
b What is the special name of this triangle?
c What is special about the sides of this triangle?

5 In the triangle on the right, two of the angles are the same.
a Work out the size of the lettered angles.
b What is the special name of this triangle?
c What is special about the sides AB and AC of this triangle?

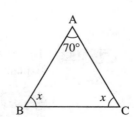

6 In the triangle on the right, the size of the angle at C is twice the size of the angle at A. Work out the size of the lettered angles.

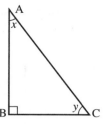

7 Find the size of the angle marked with a letter in each of the diagrams.

a

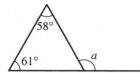

b

FM 8 The diagram shows a guy rope attached to a mast of a marquee.

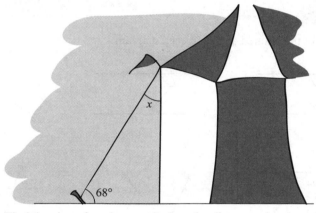

Find the size of angle *x* marked on the diagram.

AU 9 Here are five statements about triangles. Some are true and some are false.

1.   A triangle can have three acute angles.
2.   A triangle can have two acute angles and one obtuse angle.
3.   A triangle can have one acute angle and two obtuse angles.
4.   A triangle can have two acute angles and one right-angle.
5.   A triangle can have one acute angle and two right-angles.

If a statement is true, draw a sketch of a triangle.
If a statement is false, explain why.

AU 10

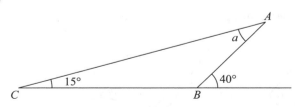

Explain why angle *a* is 25 degrees.

AU 11 A town planner has drawn this diagram to show three paths in a park, but they have missed out the angle marked *x*.

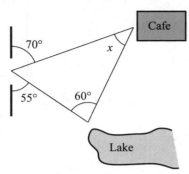

7.4 Parallel lines

HOMEWORK 7D

1 State the size of the lettered angles in each diagram.

a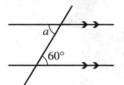

b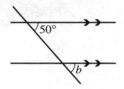

c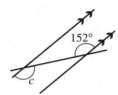

d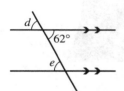

e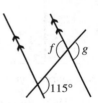

f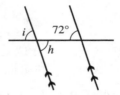

2 State the size of the lettered angles in each diagram.

a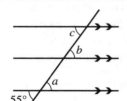

b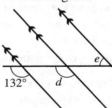

c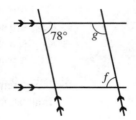

3 State the size of the lettered angles in these diagrams.

a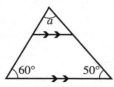

b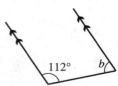

4   Find the values of *x* and *y* in these diagrams.

a

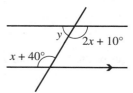

b

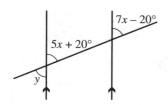

5   ABC is an isosceles triangle with angle ABC 52°.
XY is parallel to BC.

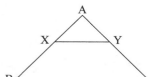

Work out the size of angle BAC.

PS   **6**   Work out the size of angle *r* in terms of *p* and *q*.

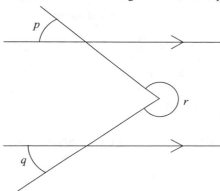

> **HINTS AND TIPS**
>
> Draw a third parallel line that passes through angle *r*.

AU   **7**   In the diagram, AB is parallel to CD.

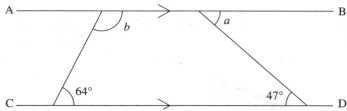

Explain why *a* + *b* = 163 degrees.

7.5 Special quadrilaterals

HOMEWORK 7E

D

1 For each of these trapeziums, calculate the sizes of the lettered angles.

a 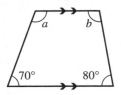 **b** 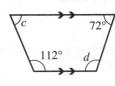 **c**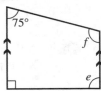

2 For each of these parallelograms, calculate the sizes of the lettered angles.

a 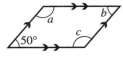 **b** 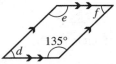 **c**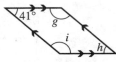

3 For each of these kites, calculate the sizes of the lettered angles.

a   **b**   **c**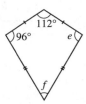

4 For each of these rhombuses, calculate the sizes of the lettered angles.

a 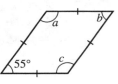 **b**   **c**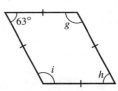

FM 5 The diagram shows the side wall of a barn. The owner wants the angle between the roof and the horizontal to be no less than 20 degrees, so that rain will run off quickly. What can you say about the size of angles A and D?

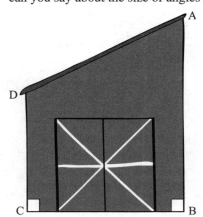

PS **6** The diagram shows a parallelogram ABCD.
AC is a diagonal.

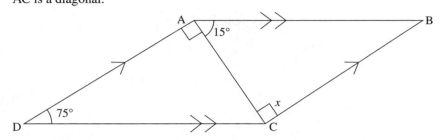

a Find angle x.
b Work out angle BCD.

AU **7** Give two reasons to explain why the trapezium is different from the parallelogram.

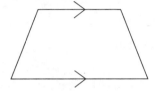

Functional Maths Activity

Back bearings

The three-figure bearing of B from A is 060°.

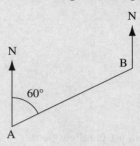

The three-figure bearing of A from B is known as a *back bearing*.

The diagram below shows that the back bearing of A from B is 240°

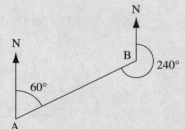

Trevor is taking an examination for his pilot's licence.
Here is one of his questions.

The map shows three airports in England.

a Use the map to find the three-figure bearing
for each of the following flights:
i Newcastle from Manchester
ii Heathrow from Newcastle
iii Manchester from Heathrow.

b Calculate the back bearing for each flight.

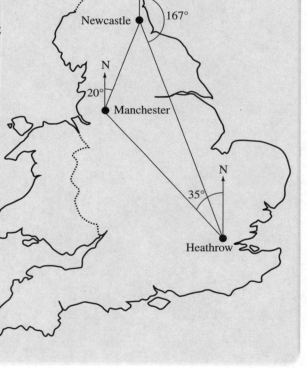

8.1 The language of algebra

HOMEWORK 8A

AU FM **1** Bill has 78p. Ben has 62p. How much will Bill have to give Ben so they both have the same amount?

2 Write down the algebraic expression that says:
- **a** 4 more than x
- **b** 7 less than x
- **c** k more than 3
- **d** t less than 8
- **e** x added to y
- **f** x multiplied by 4
- **g** 5 multiplied by t
- **h** a multiplied by b
- **i** m divided by 2
- **j** p divided by q.

3 Val is x years old. Dave is four years older than Val and Ella is five years younger than Val.
- **a** How old is Dave?
- **b** How old is Ella?

4 A packet contains n sweets.

The total number of sweets here is $2n + 3$.

Write down an expression for the total number of sweets in the following.

a

b

c

AU **5** Here is a two-step number machine.

Number IN → Add 12 → Subtract 5 → Number OUT

a Use the number machine to complete this table.

Number in	Number out
8	15
16	
	21
100	

b The number machine can be simplified. The two steps can be made into one step. What will this step be?

Number IN → Number OUT

c   The number IN is y. Write and expression for the number OUT.

Number in	Number out
y	

PS 6   My brother is five years younger than me. The total of our ages is 27 years. How old am I?

7   Sue has p pets.
- Frank has two more pets than Sue.
- Chloe has three fewer pets than Sue.
- Lizzie has twice as many pets as Sue.

How many pets does each person have?

8   **a**   Tom has £20 and spends £16. How much does he have left?
b   Sam has £10 and spends £a. How much does he have left?
c   Ian has £b and spends £c. How much does he have left?

9   **a**   How many days are there in three weeks?
b   How many days are there in z weeks?

10   **a**   Granny Parker divides £30 equally between her three grandchildren. How much does each receive?
b   Granny Smith divides £r equally between her four grandchildren. How much does each receive?
c   Granny Thomas divides £p equally between her q grandchildren. How much does each receive?

8.2 Simplifying expressions

HOMEWORK 8B

1   Evaluate these expressions, writing them as simply as possible.

a $3 \times 4t$	**b** $2 \times 5y$	**c** $4y \times 2$	**d** $3w \times 3$
e $4t \times t$	**f** $6b \times b$	**g** $3w \times w$	**h** $6y \times 2y$
i $5p \times p$	**j** $4t \times 32t$	**k** $5m \times 4m$	**l** $6t \times 4t$
m $m \times 7t$	**n** $5y \times w$	**o** $8t \times q$	**p** $n \times 69t$
q $5 \times 6q$	**r** $5f \times 2$	**s** $6 \times 3k$	**t** $5 \times 7r$

2   Evaluate these expressions, writing them as simply as possible.

a $t^2 \times t$	**b** $p \times p^2$	**c** $5m \times m^2$	**d** $3t^2 \times t$
e $4n \times 2n^2$	**f** $5r^2 \times 4r$	**g** $t^2 \times t^2$	**h** $k^3 \times k^2$
i $8n^2 \times 2n^3$	**j** $4t^3 \times 3t^4$	**k** $7a^4 \times 2a^3$	**l** $k^5 \times 3k^2$
m $-k^2 \times -k$	**n** $-5y \times -2y$	**o** $-3d^2 \times -6d$	**p** $-2p^4 \times 6p^2$
q $5mq \times q$	**r** $4my \times 3m$	**s** $4mt \times 3m$	**t** $5qp \times 2qp$

FM 3   A bee hive has 4000 bees. It is infected with a disease that kills off half the remaining bees each day. About how many bees will be left after a week?

AU 4   **a**   Which of the following is **not** equivalent to $12m^3$?

   A: $3m^2 \times 4m$ B: $2m \times 6m^2$ C: $1 \times 12m^2$ D: $m \times 12m^2$

b   Simplify the expression chosen in **a**.

PS 5 Square A has a side of $4x$ cm.
Square B has an area of $4x^2$ cm^2
Calculate the difference in the areas of the two squares.

HOMEWORK 8C

Example 1 $x + 3x + 2x - 4x = 2x$

Example 2 $2a + 3b + 5a + 2b - 4a - b = 3a + 4b$

Example 3 $2x^2 + 4x^2 - x^2 = 5x^2$

Example 4 $5x^2 + 3y - 3x^2 - 4y = 2x^2 - y$

1 Write each of these expressions in a shorter form.
  a $a + a + a$ **b** $3b + 2b$ **c** $3c + c + 5c$
  d $5d - d$ **e** $5e + 2e - 4e$ **f** $7f - 2f + 3f$
  g $2g + 4g - 6g$ **h** $4h - 6h$ **i** $3i^2 + 2i^2$
  j $5j^2 + j^2 - 2j^2$

2 Simplify each of the following expressions.
  a $2y + 5x + y + 3x$ **b** $4m + 6p - 2m + 4p$ **c** $3x + 6 + 3x - 2$
  d $7 - 5x - 2 + 8x$ **e** $5p + 2t + 3p - 2t$ **f** $4 + 2x + 4x - 6$
  g $4p - 4 - 2p - 2$ **h** $4x + 3y + 2x - 5y$ **i** $4 + 3t + p - 6t + 3 + 5p$
  j $4w - 3k - 2w - k + 4w$

3 Simplify each of the following expressions.
  a $4x + 8 - 3x + 1$ **b** $7 - 3y - 4 + 5y$ **c** $5a + 3b - a - 5b$
  d $5c - 8d - 3c + 4d$ **e** $7x + 3y + 3 + 5y - 6$ **f** $4a + 3b - 4a - b$

4 Simplify each of the following expressions.
  a $3x^2 + 8 - 2x^2 - 3$ **b** $5a^2 + 3b - 4a^2 + 2b$ **c** $k + 3k^2 - 3k + 2k^2$
  d $3c^2 + 4d - 3c^2 - 3d$ **e** $5x^2 + 3y^2 - 3x^2 + y^2$ **f** $4y^2 + 2z^2 - 6y^2 - 3z^2$

FM 5 Two of the first recorded units of measurement were the *cubit* and the *palm*.
The cubit is the distance from the finger tip to the elbow and the palm is the distance
across the hand.
Marlon measures the size of a table top using his arm and his palm.
He finds the length to be 4 cubits and 2 palms,
and the width to be 2 cubits and 3 palms.
Later he measures his cubit as 48 cm and his palm
as 9 cm.
What is the perimeter of the table top?
Give your answer in metres.

AU 6 *ABCDEF* *ABCD* is a line such that $AC = w$ cm, $AB = x$ cm and $CD = y$ cm.

a   Write an expression for the length BC.
b   Write an expression for the length AD.

PS 7 *ABCDEF* is an L-shape.
$AB = 2x$ and $DE = x$
$AF = 3x + 1$ and $EF = 3x - 1$
a   Explain why the length $CD = x - 1$.
b   Find the perimeter of the shape in terms of x.
c   If $x = 4$ cm, what is the perimeter of the shape?

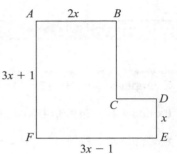

8.3 Expanding brackets

HOMEWORK 8D

AU 1 Copy the diagram below and draw lines to show which algebraic expressions are equivalent. One line has been drawn for you.

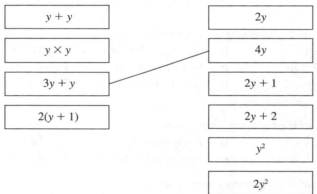

2 Expand these expressions.

a $3(4 + m)$ b $6(3 + p)$ c $4(4 - y)$ d $3(6 + 7k)$
e $4(3 - 5f)$ f $2(4 - 23w)$ g $7(g + h)$ h $4(2k + 4m)$
i $6(2d - n)$ j $t(t + 5)$ k $m(m + 4)$ l $k(k - 2)$
m $g(4g + 1)$ n $y(3y - 21)$ o $p(7 - 8p)$ p $2m(m + 5)$
q $3t(t - 2)$ r $3k(5 - k)$ s $2g(4g + 3)$ t $4h(2h - 3)$

FM 3 An approximate rule for converting degrees Fahrenheit into degrees Celsius is:
$C = 0.5(F - 30)$
a   Use this rule to convert 22°F into degrees Celsius.
b   Which of the following is an approximate rule for converting degrees Celsius into degrees Fahrenheit?
$F = 2(C + 30)$ $F = 0.5(C + 30)$ $F = 2(C + 15)$ $F = 2(C - 15)$

PS 4 The expansion $3(4x + 8y) = 12x + 24y$.
Write down two other expansions that give an answer of $12x + 24y$.

5 Expand these expressions.

a $2t(6 - 5t)$ **b** $4d(3d + 5e)$ **c** $3y(4y + 5k)$ **d** $6m^2(3m - p)$
e $y(y^2 + 7)$ **f** $h(h^3 + 9)$ **g** $k(k^2 - 4)$ **h** $t(t^2 + 3)$
i $5h(h^3 - 2)$ **j** $4g(g^3 - 3)$ **k** $5m(2m^2 + m)$ **l** $2d(4d^2 - d^3)$
m $4w(3w^2 + t)$ **n** $3a(5a^2 - b)$ **o** $2p(7p^3 - 8m)$ **p** $m^2(3 + 5m)$
q $t^3(t + 3t^2)$ **r** $g^2(4t - 3g^2)$ **s** $2t^2(7t + m)$ **t** $3h^2(4h + 5g)$

HOMEWORK 8E

1 Simplify these expressions.

a $5t + 4t$ **b** $4m + 3m$ **c** $6y + y$ **d** $2d + 3d + 5d$
e $7e - 5e$ **f** $6g - 3g$ **g** $3p - p$ **h** $5t - t$
i $t^2 + 4t^2$ **j** $5y^2 - 2y^2$ **k** $4ab + 3ab$ **l** $5a^2d - 4a^2d$

2 Expand and simplify these expressions.

a $3(2 + t) + 4(3 + t)$ **b** $6(2 + 3k) + 2(5 + 3k)$ **c** $5(2 + 4m) + 3(1 + 4m)$
d $3(4 + y) + 5(1 + 2y)$ **e** $5(2 + 3f) + 3(6 - f)$ **f** $7(2 + 5g) + 2(3 - g)$

3 Expand and simplify these expressions.

a $2(3 + h) - 3(5 + 3h)$
b $3(2g + 1) - 2(g + 5)$
c $2(3y + 2) - 3(3y + 1)$
d $4(2t + 1) - 3(3t + 1)$
e $2(5k + 3) - 3(2k - 1)$
f $4(2e + 3) - 3(3e + 2)$

4 Expand and simplify these expressions.

a $m(5 + p) + p(2 + m)$ **b** $k(4 + h) + h(5 + 2k)$ **c** $t(1 + 2n) + n(3 + 5t)$
d $p(5q + 1) + q(3p + 5)$ **e** $2h(3 + 4j) + 3j(h + 4)$ **f** $3y(4t + 5) + 2t(1 + 4y)$

FM 5 Adult tickets for a pantomime cost £x and children's tickets cost £y.
At the afternoon show there were 30 adults and 150 children.
At the evening show there were 50 adults and 120 children.

a Write down an expression for the total amount of money taken on that day in terms of x and y.
b The daily expense for putting on the show is £2500. If $x = 15$ and $y = 10$, how much profit did the theatre make that day?

AU 6 Don wrote the following:
$2(3x - 1) + 5(2x + 3) = 5x - 2 + 10x + 15 = 15x - 13$
Don has made two mistakes.
Explain the mistakes that Don has made.

PS 7 An internet site sells CDs. They cost £$(x + 0.75)$ each for the first five and then £$(x + 0.25)$ for any orders over five.

a Lo buys eight CDs. Which of the following expressions represents how much Lo will pay?

$8(x + 0.75)$ $5(x + 0.75) + 3(x + 0.25)$

$3(x + 0.75) + 5(x + 0.25)$ $8(x + 0.25)$

b If $x = 5$, how much will Lo pay?

8.4 Factorisation

HOMEWORK 8F

1 Factorise the following expressions.

a	$9m + 12t$	**b**	$9t + 6p$	**c**	$4m + 12k$	**d**	$4r + 6t$
e	$4w - 8t$	**f**	$10p - 6k$	**g**	$12h - 10k$	**h**	$2mn + 3m$
i	$4g^2 + 3g$	**j**	$4mp + 2mk$	**k**	$4bc + 6bk$	**l**	$8ab + 4ac$

2 Factorise the following expressions.

a	$3y^2 + 4y$	**b**	$5t^2 - 3t$	**c**	$3d^2 - 2d$	**d**	$6m^2 - 3mp$
e	$3p^2 + 9pt$	**f**	$8pt + 12mp$	**g**	$8ab - 6bc$	**h**	$4a^2 - 8ab$
i	$8mt - 6pt$	**j**	$20at^2 + 12at$	**k**	$4b^2c - 10bc$	**l**	$4abc + 6bed$
m	$6a^2 + 4a + 10$			**n**	$12ab + 6bc + 9bd$		
o	$6t^2 + 3t + at$			**p**	$96mt^2 - 3mt + 69m^2t$		
q	$6ab^2 + 2ab - 4a^2b$			**r**	$5pt^2 + 15pt + 5p^2t$		

3 Factorise the following expressions where possible. List those which cannot factorise.

a	$5m - 6t$	**b**	$3m + 2mp$	**c**	$t^2 - 5t$	**d**	$6pt + 5ab$
e	$8m^2 - 6mp$	**f**	$a^2 + c$	**g**	$3a^2 - 7ab$	**h**	$4ab + 5cd$
i	$7ab - 4b^2c$	**j**	$3p^2 - 4t^2$	**k**	$6m^2t + 9t^2m$	**l**	$5mt + 3pn$

FM 4 An ink cartridge is priced at £9.99.

The shop has a special offer of 20% off if you buy 5 or more.

20% of £9.99 is £1.99.

Tom wants six cartridges. Tess wants eight cartridges.

Tom writes down the calculation $6 \times 9.99 - 6 \times 1.99$ to work out how much he will have to pay.

Tess writes down the calculation $8 \times (9.99 - 1.99)$ to work out how much she will have to pay.

Both calculations are correct.

a Who has the easiest calculation and why?

b How much will they both pay for their cartridges?

AU 5 **a** Simplify these expressions.

    i $4x + 3 + 5x - 7 - 8x$

    ii $3x - 12$

    iii $x^2 - 4x$

b What do all the answers in part **a** have in common?

PS 6 A student adds up all the numbers from 1 to 100 (i.e. $1 + 2 + 3 + 4 + \ldots + 98 + 99 + 100$) using the following method:

$(1 + 100) + (2 + 99) + (3 + 98) + \ldots (50 + 51) = 50 \times 101$

a Explain why this gives the correct answer.

b What is the sum of all the numbers from 1 to 100?

8.5 Substitution

HOMEWORK 8G

Example   The expression $3x + 2$ has the value 5 when $x = 1$ and 14 when $x = 4$.

1   Find the value of $2x + 3$ when:
   a $x = 2$ **b** $x = 5$ **c** $x = 10$

2   Find the value of $3k - 4$ when:
   a $k = 2$ **b** $k = 6$ **c** $k = 12$

3   Find the value of $4 + t$ when:
   a $t = 4$ **b** $t = 20$ **c** $t = \frac{1}{2}$

4   Evaluate $10 - 2x$ when:
   a $x = 3$ **b** $x = 5$ **c** $x = 6$

5   Evaluate $5y + 10$ when:
   a $y = 5$ **b** $y = 10$ **c** $y = 15$

6   Evaluate $6d - 2$ when:
   a $d = 2$ **b** $d = 5$ **c** $d = \frac{1}{2}$

PS 7   Two of the first recorded units of measurement were the *cubit* and the *palm*.
   The cubit is the distance from the finger tip to the elbow and the palm is the distance
   across the hand.
   A cubit is four and a half palms.
   The actual length of a cubit varied throughout history but it is now accepted to be 54 cm.
   a   How many centimetres is a *palm*?
   b   Noah's Ark is recorded as being 300 cubits long by 50 cubits wide by 30 cubits high.
         What are the dimensions of the Ark in metres?

8   Find the value of $\dfrac{x + 2}{4}$ when:
   a $x = 6$ **b** $x = 10$ **c** $x = 18$

9   Find the value of $\dfrac{3x - 1}{2}$ when:
   a $x = 1$ **b** $x = 3$ **c** $x = 4$

10   Evaluate $\dfrac{20}{p}$ when:
   a $p = 2$ **b** $p = 10$ **c** $p = 20$

11   Find the value of $3(2y + 5)$ when:
   a $y = 1$ **b** $y = 3$ **c** $y = 5$

PS 12   The rule for converting degrees Fahrenheit into degrees Celsius is:
   $$C = \frac{5}{9}(F - 32)$$
   a   Use this rule to convert 68°F into degrees Celsius.
   b   Show, with a suitable substitution, that degrees F and degrees C have the same value
         at −40°.

Functional Maths Activity

Packaging

To tie up a cuboidal package that is L cm long, W cm wide and H cm high, this formula gives the length of string needed:

$S = 2L + 2W + 4H + 20$

Masood wants to send eight identical cuboidal boxes with sides of 15 cm in one package.

He can arrange them in three different ways to make a cuboidal shape.

Two of these ways are shown below.

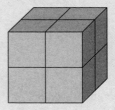 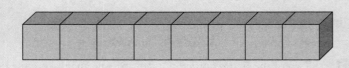

Which of the three ways of arranging the eight cubes in a cuboid will use the least amount of string? (5 marks)

9 Algebra: Graphs

9.1 Conversion graphs

FM 1 A hire firm hired out large scanners. They used the following graph to approximate what the charges would be.

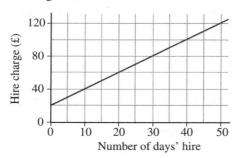

a Use the graph to find the approximate charge for hiring a scanner for:
   i 20 days   **ii** 30 days   **iii** 50 days.

b Use the graph to find out how many days' hire you would get for a cost of:
   i £120   **ii** £100   **iii** £70.

c Explain how you could use the graph to work out the cost for each day of hire.

FM 2 A conference centre used the following chart for the approximate cost of a conference based on the number of people attending it.

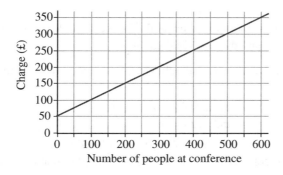

a Use the graph to find the approximate charge for:
   i 500 people   **ii** 300 people   **iii** 250 people.

b Use the graph to estimate how many people can attend a conference at the centre for a cost of:
   i £250   **ii** £150   **iii** £125.

c Explain how you could use the graph to work out the cost for each person.

FM 3 Jayne lost her fuel bill, but while talking to her friends, she found out that:

Kris who had used 750 units was charged £69
Nic who had used 250 units was charged £33
Shami who had used 500 units was charged £51.

a Plot the given information and draw a straight-line graph. Use a scale from 0 to 800 on the horizontal units axis, and from £0 to £70 on the vertical cost axis.

b Use your graph to find what Jayne will be charged for 420 units.

AU 4 This conversion graph shows the relationship between millilitres and American 'cups', which is a measure used in cooking.

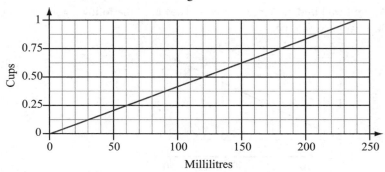

This conversion graph shows the relationship between millilitres and British fluid ounces.

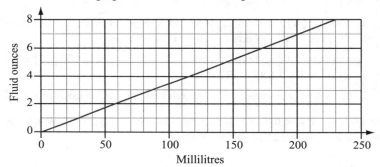

Use this information to work out how many fluid ounces would be needed for an American recipe that required one-and-a-half cups.

PS 5 Two parcel delivery companies use the following formulae for calculating delivery costs for packages:

LHD: £12.00 basic charge and £0.80 per kilometre
NTN: £10.00 basic charge and £1.00 per kilometre

LHD give a 10% discount if five or more orders are placed at the same time.
NTN give £2 off each delivery if five or more orders are placed at the same time.

Draw a conversion graph to illustrate this information.

A company needs to send five packages to five different places. The distances of each place from the company are 6, 8, 10, 12 and 15 kilometres.
Work out the cheapest way to send the packages.

9.2 Travel graphs

HOMEWORK 9B

FM 1 Joe was travelling in his car to meet his girlfriend. He set off from home at 9.00 pm, and stopped on the way for a break. This distance–time graph illustrates his journey.

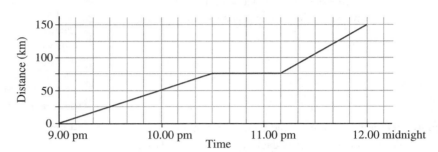

a   At what time did he:
  i   stop for his break   **ii**   set off after his break   **iii**   get to his meeting place?

b   At what average speed was he travelling:
  i   over the first hour   **ii**   over the last hour   **iii**   for the whole of his journey?

c   Joe was 10 minutes late arriving. If he had stopped for half as long, would he have been on time?

FM 2 Jean set off in a taxi from Hellaby. The taxi then went on to pick up Jeans's parents. It then travelled further, dropping them all off at a shopping centre. The taxi went on a further 10 km to pick up another party and took them back to Hellaby. This distance–time graph illustrates the journey.

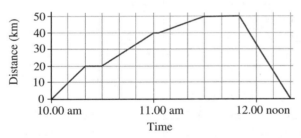

a   How far from Hellaby did Jean's parents live?
b   How far from Hellaby is the shopping centre?
c   What was the average speed of the taxi while only Jean was in the taxi?
d   What was the average speed of the taxi back to Hellaby?

FM 3 Grandad took his grandchildren out for a trip. They set off at 1.00 pm and travelled, for half an hour, away from Norwich at an average speed of 60 km/h. They stopped to look at the sea and have an ice cream. At two o'clock, they set off again, travelling for a quarter of an hour at a speed of 80 km/h. Then they stopped to play on the sand for half an hour. Grandad then drove the grandchildren back home at an average speed of 50 km/h.

Draw a travel graph to illustrate this story. Use a horizontal axis to represent time from 1.00 pm to 5.00 pm, and a vertical scale from 0 km to 60 km.

AU 4 A runner sets off at 8 am from point P to jog along a trail at a steady pace of 12 km per hour. One hour later, a cyclist sets off from P on the same trial at a steady pace of 24 km per hour. After 30 minutes, the cyclist gets a puncture which takes 30 minutes to fix. She then sets of at a steady pace of 24 km per hour.

At what time does the cyclist catch up with the runner? You may use a grid to help you solve this question.

HINTS AND TIPS

This question can be done by many methods, but doing a distance–time graph is the easiest. Mark a grid with a horizontal axis from 8 am to 12 pm and the vertical axis as distance from 0 to 40. Draw lines for both runner and cyclist. Remember that the cyclist doesn't start until 9 am.

PS 5 Three runners, A, B and C took part in a handicap race over 4 km. The graph shows how each runner ran in the race.

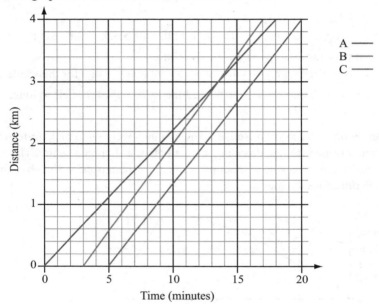

Runner A had a zero handicap and finished the race on 18 minutes.
Runner B had a 3-minute handicap and finished the race on 17 minutes.

a **i** What was runner C's handicap?

ii On what time from the start did runner C finish?

b Who ran the distance in the fastest actual time?

c The following week, the runners do the same race again. Assuming that they run at the same pace, what handicaps should all the runners have so that they finish at the same time?

9.3 Flow diagrams and graphs

HOMEWORK 9C

1 Draw the graph of $y = x + 1$

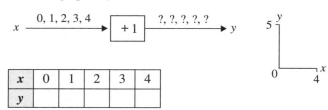

x	0	1	2	3	4
y					

2 Draw the graph of $y = 2x + 1$

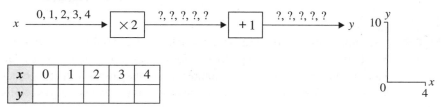

x	0	1	2	3	4
y					

3 Draw the graph of $y = 3x + 1$

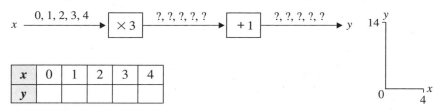

x	0	1	2	3	4
y					

4 Draw the graph of $y = x - 1$

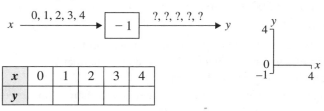

x	0	1	2	3	4
y					

5 **a** Draw the graphs of $y = x - 2$ and $y = 2x - 1$ on the same grid.
b Where do the graphs cross?

6 **a** Draw the graphs of $y = 2x$ and $y = x + 2$ on the same grid.
b Where do the graphs cross?

PS 7 This flow diagram connects two variables P and Q.

This graph also connects the variables P and Q.
Copy the graph and fill in the missing values on the P axis.

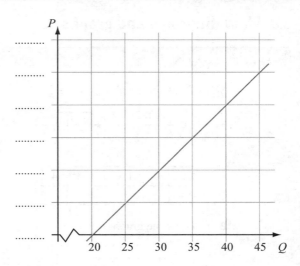

FM 8 A theatre charges £10.00 for an adult ticket and £7.50 for a child ticket.
No children are allowed to go to a performance without at least one adult present.
To work out the cost of various combinations, they use a wall chart or a flow diagram.

a Use the flow chart to work out the cost of three adults and two children.

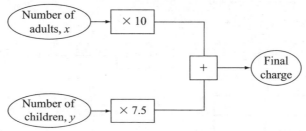

b Draw a chart to show the costs of tickets for up to five adults and four children.
Use the flow diagram to complete the chart.

c A group paid £30.00. Can you say for sure how many adults and/or children there were in the party?

AU 9 A teacher reads out the following think-of-a-number problem:
'I am thinking of a number, I divide it by 2 and then subtract 1.'

a Represent this using a flow diagram.

b If the input is x and the output is y, write down a relationship between x and y.

c Draw a graph for x values from 0 to 6.

d Explain how you could use the graph to find the number someone thought of if their final answer was 1.5.

9.4 Linear graphs

HOMEWORK 9D

Draw the graph for each of the equations given.

Follow these hints.

- Use the highest and smallest values of x given as your range.
- When the first part of the function is a division, pick x-values that divide exactly to avoid fractions.
- Always label your graphs. This is particularly important when you are drawing two graphs on the same set of axes.
- Create a table of values. You will often have to complete these in your examinations.

1 Draw the graph of $y = 2x + 3$ for x-values from 0 to 5 $(0 \leqslant x \leqslant 5)$

2 Draw the graph of $y = 3x - 1$ $(0 \leqslant x \leqslant 5)$

3 Draw the graph of $y = \dfrac{x}{2} - 2$ $(0 \leqslant x \leqslant 12)$

4 Draw the graph of $y = 2x + 1$ $(-2 \leqslant x \leqslant 2)$

5 Draw the graph of $y = \dfrac{x}{2} + 5$ $(-6 \leqslant x \leqslant 6)$

6 **a** On the same set of axes, draw the graphs of
$y = 3x - 1$ and $y = 2x + 3$ $(0 \leqslant x \leqslant 5)$
b Where do the two graphs cross?

7 **a** On the same axes, draw the graphs of
$y = 4x - 3$ and $y = 3x + 2$ $(0 \leqslant x \leqslant 6)$
b Where do the two graphs cross?

8 **a** On the same axes, draw the graphs of
$y = \dfrac{x}{2} + 1$ and $y = \dfrac{x}{3} + 2$ $(0 \leqslant x \leqslant 12)$
b Where do the two graphs cross?

9 **a** On the same axes, draw the graphs of
$y = 2x + 3$ and $y = 2x - 1$ $(0 \leqslant x \leqslant 4)$
b Do the graphs cross? If not, why not?

10 **a** Copy and complete the table to draw the graph of
$x + y = 6$ $(0 \leqslant x \leqslant 6)$

x	0	1	2	3	4	5	6
y							

b Now draw the graph of $x + y = 3$ $(0 \leqslant x \leqslant 6)$

FM 11 CityCabs uses this formula to work out the cost of a journey of k kilometres:
$C = 2.5 + k$
TownCars uses this formula to work out the cost of a journey of k kilometres:
$C = 2 + 1.25k$
a Draw a grid of distance (k) against cost (C) like the one below. Represent these formulae as lines on your grid.

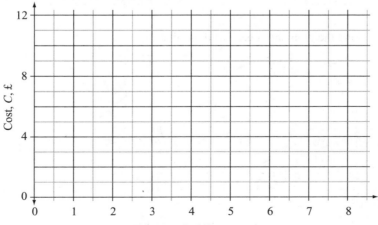

b At what length of journey do CityCabs and TownCars charge the same amount?

AU 12 The line $x + y = 5$ is drawn on the grid below.

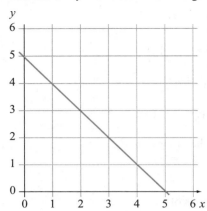

Choose values for a and b and draw the line $x = a$ and $y = b$ so that the area between the three lines is 4.5 square units.

PS 13 The two graphs shown show y against x and y against z.

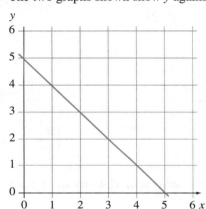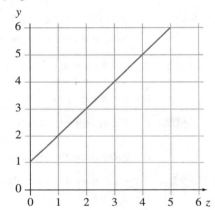

Draw a grid to show the graph of x against z.

HOMEWORK 9E

1 Find the gradient of each of these lines.

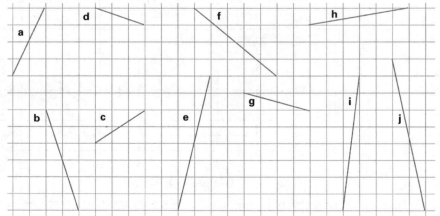

2 Draw lines with gradients of:

a 3   **b** $\frac{1}{2}$ **c** −1

d 8   **e** $\frac{3}{4}$ **f** $-\frac{1}{3}$

3 **a** Draw a pair of axes with both x and y showing from −10 to 10.

b Draw on the grid lines with the following gradients; start each line at the origin. Clearly label each line.

     i $\frac{1}{2}$ **ii** 1   **iii** 2

     iv 4   **v** −4   **vi** −2

     vii −1   **viii** $-\frac{1}{2}$

c Describe the symmetries of your diagram.

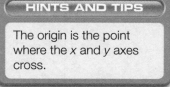

HINTS AND TIPS

The origin is the point where the x and y axes cross.

4 This graph shows the journey of a car from London to Brighton and back again. The car leaves at 8 am and returns at 3 pm.

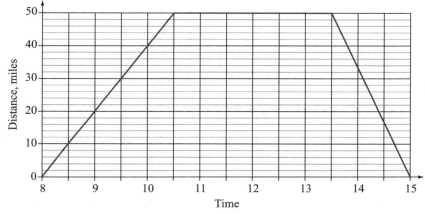

a How long does the car stop for in Brighton?

b Was the car travelling faster on the way to Brighton or on the way back to London? Explain how you can tell this from the graph.

FM PS **5** The diagram shows the safe position for a ladder. Use a grid like the one below and a protractor to work out the approximate safe angle between the ladder and the ground, marked x on the diagram.

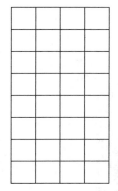

AU **6** A class is asked to predict the y-value for an x-value of 20 for this line.

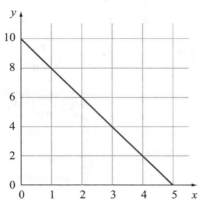

Steph says: 'The gradient is -1, so the line is $y = -x + 10$. When $x = 20$, $y = -10$.'
Steph is wrong.
Explain why and work out the correct y-value when $x = 20$.

Functional Maths Activity

Driving in the United States

a Petrol is sold in gallons in the United States and in litres in the United Kingdom.
Until recently, petrol was sold in Imperial gallons in the United Kingdom.
These two graphs show the conversions between United States gallons and Imperial
gallons and between Imperial gallons and litres.

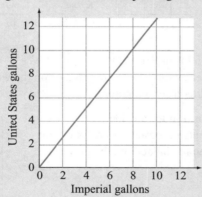 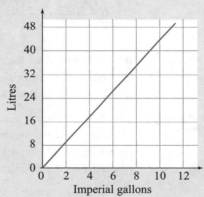

Approximately how many litres are there in a US gallon?

b This is an American advert for a car.

Dodge Journey 25 MPG Highway

Functional Maths Activity (continued)

This is a map of the highway distances between four cities in the United States.

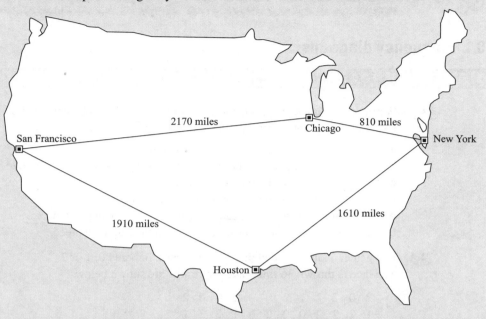

Jeff goes on a road trip from New York to Chicago, to San Francisco, to Houston and back to New York.

A US gallon of petrol costs $2.12 on average.

A litre of petrol in the UK costs £0.96 on average.

The exchange rate between pounds and dollars is £1 = $1.60

Work out:

i   How many litres of fuel would be used on this journey.

ii   How much more it would cost at UK prices than at US prices for the fuel.

10 Statistics: Statistical representation

10.1 Frequency diagrams

HOMEWORK 10A

1 For the following surveys, decide whether the data should be collected by:
i sampling **ii** observation **iii** experiment.

a The number of 'doubles' obtained when throwing two dice.
b The number of people who use a zebra-crossing on a busy main road.
c People's choice of favourite restaurant.
d The makes of cars parked in the staff car-park.
e The number of times a 'head' appears when throwing a coin.
f The type of food students prefer to eat in the school canteen.

2 In a game, a fair six-sided dice has its faces numbered 0, 1 or 2.
The dice is thrown 36 times and the results are shown below.

```
2 0 2 2 1 2 0 2 2 0 0 2
1 2 2 2 0 2 2 0 1 2 2 1
0 2 2 0 2 0 2 2 0 1 2 0
```

a Copy and complete the frequency table for the data.

Number	Tally	Frequency
0		
1		
2		

b Based on the results in the table, how many times do you think each number appears on the dice?

3 The table shows the average highest daily temperature recorded during August in 24 cities around the world.

City	Temperature (°C)	City	Temperature (°C)
Athens	33	Madras	35
Auckland	15	Marrakesh	38
Bangkok	32	Moscow	22
Budapest	27	Narvik	16
Buenos Aires	16	Nice	29
Cape Town	18	Oporto	25
Dubai	39	Perth	17
Geneva	25	Pisa	30
Istanbul	28	Quebec	23
La Paz	17	Reykjavik	14
London	20	Tokyo	30
Luxor	41	Tripoli	31

FM Functional Maths **AU** (AO2) Assessing Understanding **PS** (AO3) Problem Solving

a Copy and complete the grouped frequency table for the data.

Temperature (°C)	Tally	Frequency
11–15		
16–20		
21–25		
26–30		
31–35		
36–40		
41– 45		

b In how many cities was the temperature higher than the temperature in London?

c Kay said that the difference between the highest and lowest temperatures was 34°C but Derek said that it is was 27°C. Explain how they obtained different answers.

4 Heather attends a Spanish evening class at her local college. One evening she conducted a survey of the ages of all the people who attended. She wrote down all the ages on a piece of paper as follows.

```
25  41  33  24  46
  37  40  32  59
64  37  26  44
  58  31  29  19
  37  30  22
  48  51  68  28  27
      51  34  49
```

a How many people attended on that evening?

b Copy and complete the grouped frequency table for the data.

Age	Tally	Frequency
11–20		
21–30		
31–40		
41–50		
51–60		
61–70		

c How many people were aged under 21? Suggest a possible reason for this.

5 Pat measured the heights, to the nearest centimetre, of all the students in her class. Her data is given below.

143 135 147 153 146 138 151
142 139 131 144 127 143 145
140 143 153 141 150 137 136
125 136 140 131 147 154 142

a Draw a grouped frequency table for the data using class intervals 125–129, 130–134, 135–139, …

b In which interval do the most heights lie?

c How many students had a height of 140 cm or more?

FM 6 Mr Speed's class did a geology test with 14 questions. The test scores are shown below.

7 9 9 10 12 11 10 11 11 12 12 13 13 13
10 9 9 10 9 9 9 10 8 8 7 8 7 8 7 9

He wanted to assess how well the class had done.

a Create a frequency table showing these results.

b What test score appears the most?

c Describe the class test results.

PS 7 Azam timed how long each patient waited in a hospital's casualty department one evening. The following is his record in minutes.

28 4 31 11 31 6 24 5 36 17 10 29 7 15 20
26 32 19 27 13 24 7 15 32 8 4 38 19 34 12
33 22 34 21 25 9 25 5 18 9 34 13 23 40 8 35 20 29

Find the best way to put this data into a frequency chart to illustrate the length of time that different patients had to wait.

Explain why you chose the time ranges you used.

AU 8 Hannah wants to do a survey on the prices charged for soft drinks in her neighbourhood. She said, 'I will make a frequency table with the regions 1p–40p, 40p–60p and 60p–£1.'

Give two reasons why these class divisions may present a problem.

10.2 Statistical diagrams

HOMEWORK 10B

1 The pictogram shows the number of copies of *The Times* sold by a newsagent in a particular week.

		Total
Monday	▬ ▬ ▬	12
Tuesday	▬ ▬ ▬ ▬	
Wednesday	▬ ▬ ▬	
Thursday	▬ ▬ ▬ ▬	
Friday		
Saturday		

a How many newspapers does the symbol ▬ represent?

b Complete the totals for Tuesday, Wednesday and Thursday.

c The newsagent sold 15 copies on Friday and 22 copies on Saturday. Complete the pictogram for Friday and Saturday.

2 The pictogram shows the amount of sunshine in five English holiday resorts on one day in August.

Blackpool   Brighton   Scarborough   Skegness   Torbay
✿✿✿   ✿◖   ✿✿✿   ✿✿   ✿✿✿◖

Key: ✿ represents 3 hours.

a Write down the number of hours of sunshine for each resort.

b Great Yarmouth had $5\frac{1}{2}$ hours of sunshine on the same day. Explain why this would be difficult to show on this pictogram.

3 The pictogram shows the number of call-outs five taxi drivers had on one evening.

Brian ✪ ✪

Kontaki ✪ ✪

Robert ✪ ☾

Steve ✪ ✪ ☾

Azam ✪ ☾

Key: ✪ represents 10 call-outs.

a How many call-outs did each taxi driver have?
b Explain why the symbol used in this pictogram is not really suitable.
c Joanne had 16 call-outs on the same evening. Redraw a suitable pictogram to show the call-outs for the six taxi drivers.

4 Rachel did a survey to show the number of people in each car on their way to work on a particular morning. This is a copy of her survey sheet.

No. of people in each car	Frequency
1	30
2	19
3	12
4	5
5 or more	1

Draw a pictogram to illustrate her data.

FM 5 A form did a survey to see who received most emails in a week on their school email account.

Form Teacher	✉◲
Boys	✉ ✉ ✉ ✉◲
Girls	
Teaching Assistant	

Key: ✉ represents 20 emails

a How many emails did:
    i the form teacher receive?   **ii** the boys receive?
b Draw that part of the pictogram that will show the 110 girls' emails.
c What problem does the teaching assistant's 13 emails present?

PS 6 A survey was taken on what types of shows the parents of the students had seen at the local theatre in the last six months.

	Frequency
Musicals	128
Comedy	48
Drama	80

This information had to be put into a pictogram.
Design a pictogram to show this information with a sensible number of symbols.

AU **7** A pictogram is to be made from this frequency table.

England	405
Ireland	85
Wales	115
Scotland	325

Explain why a key of five people to a symbol is not a good idea.

10.3 Bar charts

HOMEWORK 10C

1 Linda asked a sample of people 'What is your favourite soap opera?'.
The bar chart shows their replies.

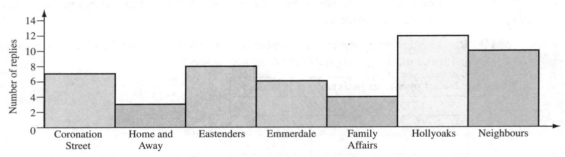

a Which soap opera got six replies?
b How many people were in Linda's sample?
c Linda collected the data from all her friends in Year 10 at school. Give two reasons
why this is not a good way to collect the data.

2 The bar chart shows the results of a survey of shoe sizes in form 10KE.

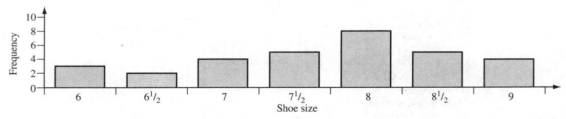

a How many students wear size $7\frac{1}{2}$ shoes?
b How many students were in the survey?
c What is the most common shoe size?
d Can you tell how many boys were in the survey? Explain your answer.

3 The table shows the lowest and highest marks six students got in a series of mental
arithmetic tests.

	Rana	Ben	Chris	Dave	Emma	Ade
Lowest mark	7	11	10	10	15	9
Highest mark	11	12	12	13	16	14

Draw a dual bar chart to illustrate the data.

4 The following data shows the times, to the nearest minute, that patients had to wait before seeing a doctor.

```
 5  12  14  24  32   7  12  35  23  27  13   6
28   4  20  13  40   5   2  11  16  31  10  26
25  30  29   9  12  27  13  20  24  11  14  38
```

a Draw a grouped frequency table to show the waiting times of the patients, using class intervals 1–10, 11–20 , 21–30, 31–40.

b Draw a bar chart to illustrate the data.

5 Richard did a survey to find out which brand of crisps his friends preferred. He drew this bar chart to illustrate his data.

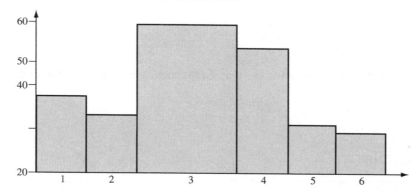

Richard's bar chart is very misleading. Explain how he could improve it and then redraw it, taking into account all your improvements.

FM 6 This table shows the number of accidents involving buses that have occurred in the town of Redlow over a six year period.

Year	2004	2005	2006	2007	2008	2009
No. of accidents	13	17	14	18	13	19

a Draw a pictogram from this data.

b Draw a bar chart from this data.

c Which diagram would you use if you were going to write an article for the local newspaper trying to show that bus drivers need to take more care? Explain your answer.

PS 7 The class was given a mental maths problem-solving test by Mr Ball. The bar chart shows the results.

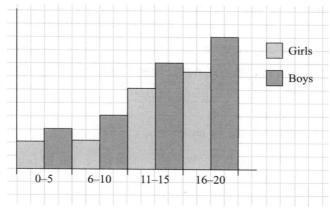

Mr Ball says, 'The boys seem to do better than the girls.'
How much better are the boys than the girls?

AU 8 The bar chart shows the rainfall of two major cities in Wales and England over three years.

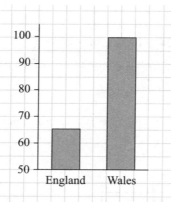

Elwyn says, 'There is more than three times as much rain in Wales than there is in England.'
Is Elwyn correct?
Explain your answer.

10.4 Line graphs

HOMEWORK 10D

1 The line graph shows the monthly average exchange rate of the Japanese Yen for £1 over a six month period.

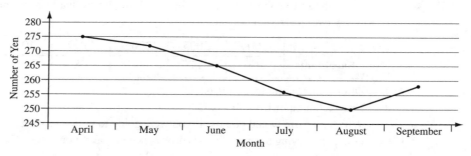

a In which month was the lowest exchange rate and what was that value?
b By how much did the exchange rate fall from April to August?
c Which month had the greatest fall in the exchange rate from the previous month?
d Mr Hargreaves changed £200 into Yen during July. How many Yen did he receive?

2 The table shows the temperature in Cuzco in Peru over a 24-hour period.

Time	0000	0400	0800	1200	1600	2000	2400
Temperature (°C)	1	−4	6	15	21	9	−1

a Draw a line graph for the data.
b From your graph estimate the temperature at 1800.

3 The table shows the value, to the nearest million pounds, of a country's imports and exports.

Year	2004	2005	2006	2007	2008	2009
Imports	22	35	48	51	62	55
Exports	35	41	56	53	63	58

a Draw line graphs on the same axes to show the imports and exports of the country.
b Find the smallest and greatest difference between the imports and exports.

FM 4 The table shows the estimated number of passengers travelling by train in a country.

Year	1970	1975	1980	1985	1990	1995	2000	2005
Train passengers (thousands)	210	310	450	570	590	650	690	770

a Draw a line graph showing this data.
b From your graph, estimate the number of passengers in 2010.
c Between which two consecutive years did the number of train passengers increase the most?
d Explain the trend in the numbers of train passengers. What reason can you suggest to explain this trend?

PS 5 The height of a baby giraffe is measured at the end of each week.

Week	1	2	3	4	5
Height (cm)	110	160	200	220	235

Estimate the height of the giraffe after 6 weeks.

AU 6 When plotting a graph to show the number of people attending cricket matches at Headingley, Kevin decided to start his graph at 18 000.
Explain why he might have done that.

10.5 Stem-and-leaf diagrams

HOMEWORK 10E

1 The following stem-and-leaf diagram shows the number of TVs a retailer sold daily over a three week period.

```
1 | 2 8 9              Key 1 | 2 represents 12 TVs
2 | 0 2 4 4 4 4 5 7 8 8 9
3 | 1 2 4 8
```

a What is the greatest number of TVs the retailer sold in one day?
b What is the most common number of TVs sold daily?
c What is the difference between the greatest number and the least number of TVs sold?

2 The following stem-and-leaf diagram shows the ages of a group of people waiting for a train at a station.

1	6 8 9
2	4 7 8 9
3	0 2 4 5 6
4	2 5 5 6 8
5	0 4 8

Key 1 | 6 represents an age of 16

a How many people were waiting for a train?
b What is the age of the youngest person?
c What is the difference in age between the youngest person and oldest person?

3 A survey is carried out to find the speed, in miles per hour, of 30 vehicles travelling on a motorway. The results are shown below.

62 45 70 58 68 70 75 80 72 65 40 55 65 72 38
70 75 68 50 48 65 60 68 72 70 45 68 69 68 60

a Show the data on an ordered stem-and-leaf diagram. (Remember to show a key.)
b What is the most common speed?
c What is the difference between the greatest speed and the lowest speed?

FM 4 A teacher has given her class a spelling test. She wants to compare the results of boys and girls. To do this she puts the results in a stem-and-leaf table.

	Boys		Girls	
5 3 1 2	1	4 6 8		
8 8 5 1	2	1 1 2 7 7 7		
6 5 5 5	3	0 0 4		

Key: Boys 2 | 1 means 12 correct
Girls 1 | 4 means 14 correct

a What was the highest number correct for the boys?
b What was the lowest number correct for the girls?
c What was the most common correct number correct for:
  i Boys   **ii** Girls?
d Overall, who did better in the tests – boys or girls?
Give a reason for your answer.

PS 5 The following data is gathered about the girls' and boys' IQ test scores in a form.
Boys: 117 105 128 132 110 108 123 114 128 110
Girls: 122 137 113 118 131 129 104 120 117 134
Present these scores in an appropriate way for them to be compared.

AU 6 Godwin was asked to create a stem-and-leaf diagram from some numerical data, but he said, 'It is impossible to do this sensibly!'
Give an example of 10 items of numerical data that could not sensibly be put into a stem-and-leaf diagram.

Functional Maths Activity

Wine buying in the UK

This table shows the number of cases of different types of wine bought in the UK over the last few years.

	Number of cases (millions)				
Wines	**2005**	**2006**	**2007**	**2008**	**2009**
Chardonnay	6	8	8	8	8
Pinot Grigio	2	3	4	5	5
Sauvignon Blanc	3	4	5	5	6
Merlot	4	4	5	4	5
Shiraz	3	3	4	5	5
Cabernet Sauvignon	4	5	4	5	5
Rioja	2	2	2	2	2

Use appropriate statistical diagrams and measures to summarise the data given in the table.
Then write a report about the sales of wines in the UK over these five years.

Statistics: Averages

11.1 The mode

HOMEWORK 11A

Example   Terry scored the following number of goals in 12 school football matches:

1 2 1 0 1 0 0 1 2 1 0 2

The number which occurs most often in this list is 1. So, the mode is 1.
We can also say that the modal score is 1.

G

1 Find the mode for each set of data.
  a 3, 1, 2, 5, 6, 4, 1, 5, 1, 3, 6, 1, 4, 2, 3, 2, 4, 2, 4, 2, 6, 5
  b 17, 11, 12, 15, 11, 13, 18, 14, 17, 15, 13, 15, 16, 14
  c 110, 10, 101, 10, 111, 110, 11, 101, 11, 111, 11, 101, 101, 111
  d 1, −3, 3, 2, −1, 1, −3, −2, 3, −1, 2, 1, −1, 1, 2
  e 7, $6\frac{1}{2}$, 6, $7\frac{1}{2}$, 8, $5\frac{1}{2}$, $6\frac{1}{2}$, 6, 7, $6\frac{1}{2}$, 7, $6\frac{1}{2}$, 6, $7\frac{1}{2}$

2 Find the modal category for each set of data.
  a I, A, E, U, A, O, A, E, U, A, I, A, E, I, E, O, E, I, E, O
  b ITV, C4, BBC1, C5, BBC2, C4, BBC1, C5, ITV, C4, BBC1, C4, ITV
  c ↑, →, ↑, ←, ↓, →, ←, ↑, ←, →, ↓, ←, ←, ↑, →, ↓
  d ♥, ♣, ♦, ♣, ♠, ♥, ♣, ♦, ♣, ♦, ♥, ♠
  e ¥, €, £, €, $, £, ¥, €, £, $, €, £, $, €

3 Farmer Giles kept a record of the number of eggs his hens laid. His data is shown on the diagram below.

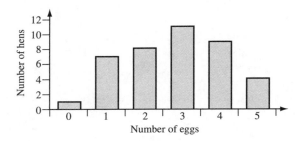

  a How many hens did Farmer Giles have?
  b What is the mode of the number of eggs laid?
  c How many eggs were laid altogether?

PS 4 What is the mode of these numbers?
  5 7 1 3 7 8 7 9 2 3 1 6
  1 3 5 6 4 2 7 8 9 2 9 4
  5 1 4 6 8 3 9 3 2 2 4 5
  7 5 8 7 3 7 8 2 3 5 6 6
  3 4 3 1 2 3 5 6 5 4 4

FM 5 This table shows favourite pets for students in form Y10E.

	Dog	Cat	Rabbit	Guinea pig	Other
Boys	8	1	3	2	0
Girls	3	3	7	3	1

   a How many students are in the form Y10E?

   b What is the modal pet for:

     i boys   **ii** girls   **iii** the whole form?

   c After two new students join the form, the modal pet for the form is rabbit.
     Which of the following statements are true?

     1 Both students like cats.

     2 Both students like rabbits.

     3 One student likes rabbits; the other likes cats.

     4 You cannot tell what pets they like.

AU 6 The makes of cars in a supermarket car park were noted one day as:

Ford, Peugeot, Austin, Honda, Ford, Austin, Peugeot, Ford, Peugeot, Austin, Ford, Toyota, Honda, Honda, Peugeot, Ford, Honda, Austin, Honda, Peugeot, Toyota, Toyota, Toyota, Austin, Toyota.

What is the problem with finding the modal make of car in this car park at this time?

7 The grouped frequency table shows the number of marks a class of students obtained in a spelling test out of 30 marks.

Number of marks	Frequency
1–5	1
6–10	2
11–15	4
16–20	8
21–25	10
26–30	5

   a How many pupils are in the class?

   b Write down the modal class for the number of marks.

   c Stan looked at the table and said that at least one person got full marks. Explain why he may be wrong.

8 The data shows the times, to the nearest minute, that 30 shoppers had to wait in the queue at a checkout of a supermarket.

     1   3   8   12   7   4   0   9   10   15   8   1   2   7   4

     2   4   7   1   0   5   4   8   4   10   7   5   4   1   5

   a Copy and complete the grouped frequency table.

Time in minutes	Tally	Frequency
0–3		
4–7		
8–11		
12–15		

   b Draw a bar chart to illustrate the data.

   c How many shoppers had to wait more than seven minutes?

   d Write down the modal class for the time that the shoppers had to wait?

   e How could the supermarket manager decrease the waiting time of the shoppers?

11.2 The median

HOMEWORK 11B

The median is the value at the middle of a list of values after they have been put in order of size, from lowest to highest.

Example 1	Find the median for the list of numbers below.

2, 3, 5, 6, 1, 2, 3, 4, 5, 4, 6
Putting the list in numerical order gives
1, 2, 2, 3, 3, 4, 4, 5, 5, 6, 6
There are 11 numbers in the list, so the middle of the list is the sixth number. Therefore, the median is 4.

Example 2	The ages of 20 people attending a conference are shown below.

28, 32, 46, 23, 28, 34, 52, 61, 45, 34, 39, 50, 26, 44, 60, 53, 31, 25, 37, 48
Draw a stem-and-leaf diagram and hence find the median age of the group.
Taking the tens to be the 'stem' and the units to be the 'leaves', the stem-and-leaf diagram is shown below. (**2 | 3** means 23)

```
2 | 3 5 6 8 8
3 | 1 2 4 4 7 9
4 | 4 5 6 8
5 | 0 2 3
6 | 0 1
```

There is an even number of values in this list, so the middle of the list is between the two central values, 37 and 39. Therefore, the median is the value which is exactly midway between 37 and 39. Hence, the median is 38.

1 Find the median for each set of data.
   a 18, 12, 15, 19, 13, 16, 10, 14, 17, 20, 11
   b 22, 28, 42, 37, 26, 51, 30, 34, 43
   c 1, −3, 0, 2, −4, 3, −1, 2, 0, 1, −2
   d 12, 4, 16, 12, 14, 8, 10, 4, 6, 14
   e 1.7, 2.1, 1.1, 2.7, 1.3, 0.9, 1.5, 1.8, 2.3, 1.4

2 The weights of eleven men in a local rugby team are shown below.
   81 kg, 85 kg, 82 kg, 71 kg, 62 kg, 63 kg, 62 kg, 64 kg, 70 kg, 87 kg, 74 kg
   a Find the median of their weights.
   b Find the mode of their weights.
   c Which is the better average to use? Explain your answer.

3 The bar chart shows the scores obtained in 20 throws of a dice.
   a Write down the modal score.
   b Find the median score. Remember you must take into account all the scores.
   c Do you think that the dice is biased? Explain your answer.

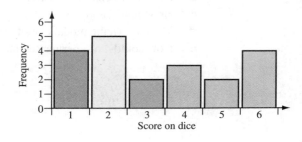

FM 4 Mrs Roberts has given her class a spelling test. She wants to compare the results of boys and girls. To do this she puts the results in a stem-and-leaf table.

Girls					Boys						
5	3	1	**1**	4	6	9	9				
8	8	5	1	**2**	1	2	2	7	7	7	
9	5	5	5	**3**	2	3					

Key: 1 | 4 represents 14 marks for boys
5 | 2 represents 25 marks for girls

a What was the modal mark for the boys?
b What was the modal mark for the girls?
c What was the median mark for the boys?
d What was the median mark for the girls?
e Who did better in the test – the boys or the girls?
Give a reason for your choice.

PS 5 Look at this list of numbers.

3   3   4   7   9   10   11   14   14   15   19

a Add four numbers to the list to make the median 11.
b Add six numbers to the list to make the median 11.
c What is the least number of numbers you can add to the list to make the median 3?

AU 6 Explain why the median is not a good average to use in this set of weights.

5 g   7 g   10 g   200 g   4 kg

7 **a** Write down a list of seven numbers that has a median of 10 and a mode of 20.
b Write down a list of eight numbers that has a median of 10 and a mode of 20.

8 The marks of 21 students in a Science modular test are shown below.

45, 62, 27, 77, 40, 55, 80, 87, 49, 57, 35, 52, 59, 78, 48, 67, 43, 68, 38, 72, 81

Draw a stem-and-leaf diagram to find the median.

FM 9 Jack is doing his Statistics project based on 'Pulse Rates'. One of his hypotheses is that a person's pulse rate increases after exercise. To test his hypothesis he asks 31 students in a PE lesson to take their pulse rate at the start of the lesson and again at the end of the lesson. He records the data on an observation sheet and then illustrates it on a back-to-back stem-and-leaf diagram.

Before exercise									Stem	After exercise					
			9	9	8	5	3		**5**						
		8	8	5	4	3	2	1	**6**						
9	8	7	7	4	2	2	2	0	**7**	1	1	2			
				9	6	5	2	2	**8**	0	0	2	7	8	
					8	7	5		**9**	0	2	2	3	5	5	8
							4	2	**10**	1	2	4	5	8	9
									11	4	6	8	9		
									12	2	5	6			
									13	8	9				
									14	4					

(**7** | 2 means 72 beats per minute)

a Find the median pulse rate of the students at the start of the lesson.
b Find the median pulse rate of the students at the end of the lesson.
c What conclusions can Jack draw from the stem-and-leaf diagram?

11.3 The mean

HOMEWORK 11C

The mean of a set of data is the sum of all the values in the set divided by the total number of values in the set.

That is, mean = $\dfrac{\text{Sum of all values}}{\text{Total number of values}}$

> **Example**   Find the mean of 4, 8, 7, 5, 9, 4, 8, 3.
>
> Sum of all the values = 4 + 8 + 7 + 5 + 9 + 4 + 8 + 3 = 48
> Total number of values = 8
> Therefore, mean = 48 ÷ 8 = 6

1   Find the mean for each set of data.
   a   4, 2, 5, 8, 6, 4, 2, 3, 5, 1
   b   21, 25, 27, 20, 23, 26, 28, 22
   c   324, 423, 342, 234, 432, 243
   d   2.5, 3.6, 3.1, 4.2, 3.5, 2.9
   e   1, 4, 3, 0, 1, 2, 5, 0, 2, 4, 2, 0

2   Calculate the mean for each set of data, giving your answer correct to one decimal place.
   a   17, 24, 18, 32, 16, 28, 20
   b   92, 101, 98, 102, 95, 104, 99, 96, 103
   c   9.8, 9.3, 10.1, 8.7, 11.8, 10.5, 8.5
   d   202, 212, 220, 102, 112, 201, 222
   e   4, 2, −1, 0, 1, −3, 5, 0, −1, 4, −2, 1

3   A group of eight people took part in a marathon to raise money for charity. Their times to run the marathon were:
   2 hours 40 minutes, 3 hours 6 minutes, 2 hours 50 minutes, 3 hours 25 minutes,
   4 hours 32 minutes, 3 hours 47 minutes, 2 hours 46 minutes, 3 hours 18 minutes
Calculate their mean time in hours and minutes.

4   The monthly wages of 11 full-time staff who work in a restaurant are as follows:
   £820, £520, £860, £2000, £800, £1600, £760, £810, £620, £570, £650
   a   Find their median wage.
   b   Calculate their mean wage.
   c   How many of the staff earn more than:
      i the median wage   **ii** the mean wage?
   d   Which is the better average to use? Give a reason for your answer.

AU 5   Town scored 30 goals in ten games of football. This means that they have an average of three goals per game.
What is the smallest number of goals they need to score in their next match to get a higher average score?

FM 6 The table shows the marks that five couples obtained in a dancing competition.

	Kath & Brian	Tom & Helen	Joe & Nik	Azan & Phyllis	David & Hannah
Tango	10	6	4	8	6
Salsa	6	8	3	8	6
Ballroom	8	4	4	8	8

a Kath said that the Salsa was the hardest dance. Find the mean score for each dance and use these to decide whether Kath is correct.

b Which pair obtained the score closest to the mean in all dances?

c How many pairs were above average in all these dances?

PS 7 Two families took part in a tug o' war competition.

Key family	Charlton family
Brian weighed 58 kg	David weighed 60 kg
Ann weighed 32 kg	Hannah weighed 56 kg
Steve weighed 49 kg	Pete weighed 42 kg
Alison weighed 39 kg	Barbara weighed 76 kg
Jill weighed 64 kg	Chris weighed 71 kg
Holly weighed 75 kg	Julie weighed 39 kg
Albert weighed 52 kg	George weighed 22 kg

Each family had to choose 4 members with a mean weight of between 45 and 50 kg. Choose two teams, one from each family, that have this mean weight between 45 and 50 kg.

8 The table shows the percentage marks which 10 students obtained in Paper 1 and Paper 2 of their GCSE Mathematics examination.

Ann	Bridget	Carole	Daniel	Edwin	Fay	George	Hannah	Imman	Joseph
72	61	43	92	56	62	73	56	38	67
81	57	49	85	62	61	70	66	48	51

a Calculate the mean mark for Paper 1.

b Calculate the mean mark for Paper 2.

c Which student obtained marks closest to the mean on both papers?

d How many students were above the mean mark on both papers?

9 The numbers of runs that a cricketer scored in seven innings were:
48, 32, 0, 62, 11, 21, 43

a Calculate the mean number of runs in the seven innings.

b After eight innings his mean score increased to 33 runs per innings. How many runs did he score in his eighth innings?

11.4 The range

HOMEWORK 11D

The range for a set of data is the highest value in the set minus the lowest value in the set.

> **Example**
> Rachel's marks in ten mental arithmetic tests were 4, 4, 7, 6, 6, 5, 7, 6, 9, 6.
> Her mean mark is $60 \div 10 = 6$ marks, and her range is $9 - 4 = 5$ marks.
> Robert's marks in the same tests were 6, 7, 6, 8, 5, 6, 5, 6, 5, 6.
> His mean mark is $60 \div 10 = 6$ marks, and his range is $8 - 5 = 3$ marks.
> Although the means are the same, Robert has a smaller range. This shows that
> Robert's results are more consistent.

1 Find the range for each set of data.
- **a** 23, 18, 27, 14, 25, 19, 20, 26, 17, 24
- **b** 92, 89, 101, 96, 100, 96, 102, 88, 99, 95
- **c** 14, 30, 44, 25, 36, 27, 15, 42, 27, 12, 40, 31, 34, 24
- **d** 3.2, 4.8, 5.7, 3.1, 3.8, 4.9, 5.8, 3.5, 5.6, 3.7
- **e** 5, −4, 0, 2, −5, −1, 4, −3, 2, 2, 0, 1, −4, 0, −2

2 The table shows the ages of a group of students on an 'Outward Bound' course at a Youth Hostel.

Age	14	15	16	17	18	19
Number of students	2	3	8	5	6	1

- **a** How many students were on the course?
- **b** Write down the modal age of the students.
- **c** What is the range of their ages?
- **d** Draw a bar chart to illustrate the data.

3 A travel brochure shows the average monthly temperatures, in °C, for the island of Crete.

Month	April	May	June	July	August	September	October
Temperature °C	68	74	78	83	82	75	72

- **a** Calculate the mean of these temperatures.
- **b** Write down the range of these temperatures.
- **c** The mean temperature for the island of Corfu was 77°C and the range was 20°C. Compare the temperatures for the two islands.

4 The table shows the daily attendance of three forms of 30 students over a week.

	Monday	Tuesday	Wednesday	Thursday	Friday
Form 10KG	25	25	26	27	27
Form 10RH	22	23	30	26	24
Form 10PB	24	29	28	25	29

- **a** Calculate the mean attendance for each form.
- **b** Write down the range for the attendance of each form.
- **c** Which form had **i** the best attendance and **ii** the most consistent attendance? Give reasons for your answers.

FM 5 Over a three-week period, a sandwich shop had the following takings. The owner wants to see how takings vary from week to week.

	Monday	Tuesday	Wednesday	Thursday	Friday
Week 1	£640	£585	£726	£607	£563
Week 2	£682	£661	£508	£567	£402
Week 3	£709	£682	£624	£668	£648

a Calculate the mean amount taken each week.
b Find the range for each week.
c What can you say about the amounts taken over the three weeks?

PS 6 Look at this list of children, showing their present ages and weights.

Name	Age (years)	Weight (kg)
Olly	12	24
Elinor	7	14
Latham	11	18
Aimee	13	23
Chelsie	6	13
Kemunto	7	16
Kai-Yan	6	15
Anna	5	13
Zoe	12	17
Kesia	10	16

a Find an age range of 5 that includes as many children as possible.
b Find the smallest range of weights that includes 5 children.

AU 7 The age range of a football team is 1 year.
What type of football team would you most likely expect this to be?
Explain your answer.

8 The back-to-back stem-and-leaf diagram shows the marks of 30 pupils in one of their English tests.

```
        Boys       Girls

                |  1  |  8
      8 7 6 4   |  2  |  4 5 7 9
    9 7 7 4 1 0 |  3  |  1 4 7 8
      6 5 2 0   |  4  |  0 1 3 4 7
            2   |  5  |  4
```

(4 | 2 means 42 marks)

a Find the median mark for the boys and for the girls.
b Write down the range of the marks for the boys and for the girls.
c Compare the results of the boys and the girls.

11.5 Which average to use

HOMEWORK 11E

1 **a** For each set of data find the mode, the median and the mean.
  i 6, 4, 5, 6, 2, 3, 2, 4, 5, 6, 1
  ii 14, 15, 15, 16, 15, 15, 14, 16, 15, 16, 15
  iii 31, 34, 33, 32, 46, 29, 30, 32, 31, 32, 33
b For each set of data decide which average is the best one to use and give a reason.

2 A supermarket sells oranges in bags of ten.
The weights of each orange in a selected bag are shown below.
  134 g, 135 g, 142 g, 153 g, 156 g, 132 g, 135 g, 140 g, 148 g, 155 g
a Find the mode, the median and the mean for the weight of the oranges.
b The supermarket wanted to state the average weight on each bag they sold. Which of the three averages would you advise the supermarket to use? Explain why.

FM 3 Three players were hoping to be chosen for the hockey team.
The following table shows the goals they scored in each of the last few games they played.

Adam	4, 2, 3, 2, 3, 2, 2
Faisal	4, 2, 4, 6, 2
Maya	4, 0, 4, 0, 1

The teacher said they would be chosen by their best average score.
Which average would each prefer to be chosen by?

4 The weights, in kilograms, of a school football team are shown below.
  68, 72, 74, 68, 71, 78, 53, 67, 72, 77, 70
a Find the median weight of the team.
b Find the mean weight of the team.
c Which average is the better one to use? Explain why.

5 Jez is a member of a pub quiz team and, in the last eight games, his total points are shown below.
  62, 58, 24, 47, 64, 52, 60, 65
a Find the median for the number of points he scored over the eight games.
b Find the mean for the number of points he scored over the eight games.
c The team captain wanted to know the average for each member of the team. Which average would Jez use? Give a reason for your answer.

PS 6 **a** Find three numbers that have **all** of these properties:
  i a range of 3
  ii a mean of 3
b Find three numbers that have **all** of these properties:
  i a range of 3
  ii a median of 3
  iii a mean of 3

AU 7 'What is the average score for the test?'
The teacher said, '32'.
A student said, '28'.
They were both correct.
Explain how this could be.

11.6 Frequency tables

HOMEWORK 11F

1 Find **i** the mode, **ii** the median and **iii** the mean from each frequency table below.

a A survey of the collar sizes of all the male staff in a school gave these results.

Collar size	12	13	14	15	16	17	18
Number of staff	1	3	12	21	22	8	1

b A survey of the number of TVs in pupils' homes gave these results.

Number of TVs	1	2	3	4	5	6	7
Frequency	12	17	30	71	96	74	25

2 A survey of the number of pets in each family of a school gave these results.

Number of pets	0	1	2	3	4	5
Frequency	28	114	108	16	15	8

a Each child at the school is shown in the data, how many children are at the school?
b Calculate the median number of pets in a family.
c How many families have less than the median number of pets?
d Calculate the mean number of pets in a family. Give your answer to 1 decimal place.

FM 3 A survey of the number of television sets in each family home in one school year gave these results:

No. of TVs	0	1	2	3	4	5
Frequency	1	5	36	86	72	56

a How many students are in that school year?
b Calculate the mean number of TVs in a home for this school year.
c How many homes have this mean number of TVs (if you round the mean to the nearest whole number)?
d How many homes could consider themselves average from this survey?

PS 4 A coffee stain removed four numbers (in two columns) from the following frequency table of eggs laid by 20 hens one day.

Eggs	0	1	2			5
Frequency	2	3	4			1

The mean number of eggs laid was 2.5.
What could the missing four numbers be?

AU 5 Twinkle travelled to Manchester on many days throughout the year.
The table shows how many days she travelled in each week.

Days	0	1	2	3	4	5
Frequency (no. of weeks)	17	2	4	13	15	1

Explain how you would find the median number of days that Twinkle travelled in a week to Manchester.

11.7 Grouped data

1 Find for each table of values given below:
   i the modal group   and **ii** an estimate for the mean.

a

Score	0 – 20	21 – 40	41 – 60	61 – 80	81 – 100
Frequency	9	13	21	34	17

b

Cost (£)	0.00 – 10.00	10.01 – 20.00	20.01 – 30.00	30.01 – 40.00	40.01 – 60.00
Frequency	9	17	27	21	14

FM 2 A hospital has to report the average waiting time for patients in the Accident and Emergency department. A survey was made to see how long casualty patients had to wait before seeing a doctor.
The following table summarises the results for one shift.

Time (minutes)	0 – 10	11 – 20	21 – 30	31 – 40	41 – 50	51 – 60	61 – 70
Frequency	1	12	24	15	13	9	5

  a How many patients were seen by a doctor in the survey of this shift?
  b Estimate the mean waiting time taken per patient.
  c Which average would the hospital use for the average waiting time?

PS 3 The table shows the runs scored by all the batsmen in a cricket competition.

Runs	0 – 9	10 – 19	20 – 29	30 – 39	40 – 49
Frequency	8	5	10	5	2

Helen noticed that two numbers were in the wrong part of the table and that this made a difference of 1.7 to the arithmetic mean.
Which two numbers were the wrong way round?

AU 4 The profit made each week by a charity shop is shown in the table below;

Profit	£0 – £500	£501 – £1000	£1001 – £1500	£1501 – £2000
Frequency	15	26	8	3

Explain how you would estimate the mean profit made each week.

11.8 Frequency polygons

1 The table shows the number of goals scored by a football team in 20 matches.

Goals	0	1	2	3	4
Frequency	5	7	4	3	1

  a Draw a frequency polygon to illustrate the data.
  b Calculate the mean number of goals scored per game.

2 The table shows the times taken by 50 pupils to complete a multiplication square.

Time, s, seconds	$10 < s \leqslant 20$	$20 < s \leqslant 30$	$30 < s \leqslant 40$	$40 < s \leqslant 50$	$50 < s \leqslant 60$
Frequency	4	10	16	12	8

a Draw a frequency polygon to illustrate the data.
b Calculate an estimate for the mean time taken by the pupils.

FM 3 A supermarket manager needs to check that the average waiting time for customers is no more than 5 minutes. The results of a survey of waiting times for customers at a supermarket checkout are shown in the table.

Time, m, minutes	$0 < m \leqslant 2$	$2 < m \leqslant 4$	$4 < m \leqslant 6$	$6 < m \leqslant 8$	$8 < m \leqslant 10$
Frequency	3	5	10	8	4

a Draw a frequency polygon to illustrate the data.
b Calculate an estimate for the mean waiting time for the customers.
c What advice would you give the manager about average waiting times?

PS 4 The frequency polygon shows the length of time that students spent on homework one weekend.
Calculate an estimate of the mean time spent on homework by the students.

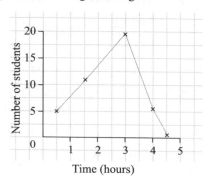

AU 5 The frequency polygon shows the times that a number of people waited at a post office before being served one morning.

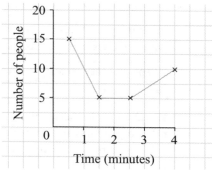

Julie said, 'Most people spent 30 seconds waiting.'
Explain why this might be wrong.

Functional Maths Activity

Words and books

1 Choose a book that doesn't have too many pictures in it and estimate how many words are in it.
You need to explain fully how you have estimated the number of words.

2 Now choose a book that does have some pictures in and estimate how many words are in this book.
Describe in detail how you have done this.
Why do you think it's so important for an editor to know roughly how many words will be in a book?

12.1 Pie charts

HOMEWORK 12A

1 The table shows the time taken by 60 people to travel to work.

Time in minutes	10 or less	Between 10 and 30	30 or more
Frequency	8	19	33

Draw a pie chart to illustrate the data.

2 The table shows the number of GCSE passes that 180 students obtained.

GCSE passes	9 or more	7 or 8	5 or 6	4 or less
Frequency	20	100	50	10

Draw a pie chart to illustrate the data.

3 Tom is doing a statistics project on the use of computers. He decides to do a survey to find out the main use of computers by 36 of his school friends. His results are shown in the table.

Main use	e-mail	Internet	Word processing	Games
Frequency	5	13	3	15

  a Draw a pie chart to illustrate his data.
  b What conclusions can you draw from his data?
  c Give reasons why Tom's data is not really suitable for his project.

4 In a survey, a TV researcher asks 120 people at a leisure centre to name their favourite type of television programme. The results are shown in the table.

Type of programme	Comedy	Drama	Films	Soaps	Sport
Frequency	18	11	21	26	44

  a Draw a pie chart to illustrate the data.
  b Do you think the sample chosen by the researcher is representative of the population? Give a reason for your answer.

FM 5 Marion is writing an article on health for a magazine. She asked a sample of people the question: 'When planning your diet, do you consider your health?' The pie chart shows the results of her question.

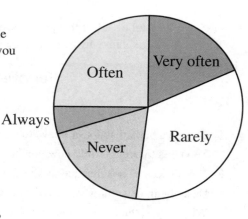

  a What percentage of the sample responded 'often'.
  b What response was given by about a third of the sample?
  c Can you tell how many people there were in the sample? Give a reason for your answer.
  d What other questions could Marion ask?

PS **6** A nationwide survey was taken on where people thought the friendliest people were in England.
What is the probability that a person picked at random from this survey answered 'East'?

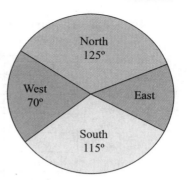

AU **7** You are asked to draw a pie chart representing the different breakfasts that students have in a morning.
What data would you need to obtain in order to do this?

12.2 Scatter diagrams

HOMEWORK 12B

1 The table below shows the heights and weights of twelve students in a class.

Student	Weight (Kg)	Height (cm)
Ann	51	123
Bridget	58	125
Ciri	57.5	127
Di	62	128
Emma	59.5	129
Flo	65	129
Gill	65	133
Hanna	65.5	135
Ivy	71	137
Joy	75.5	140
Keri	70	143
Laura	78	145

a Plot the data on a scatter diagram.
b Draw the line of best fit.
c Jayne was absent from the class, but she knows she is 132 cm tall. Use the line of best fit to estimate her weight.
d A new girl joined the class who weighed 55 kg. What height would you expect her to be?

FM 2 The table below shows the marks for ten pupils in their mathematics and music examinations.

Pupil	Maths	Music
Alex	52	50
Ben	42	52
Chris	65	60
Don	60	59
Ellie	77	61
Fan	83	74
Gary	78	64
Hazel	87	68
Irene	29	26
Jez	53	45

a Plot the data on a scatter diagram. Take the x-axis for the mathematics scores and mark it from 20 to 100. Take the y-axis for the music scores and mark it from 20 to 100.

b Draw the line of best fit.

c One of the pupils was ill when they took the mathematics examination. Which pupil was it most likely to be?

d Another pupil, Kris, was absent for the music examination but scored 45 in mathematics, what mark would you expect him to have got in music?

e Another pupil, Lex, was absent for the mathematics examination but scored 78 in music, what mark would you expect him to have got in mathematics?

PS 3 Twelve students took part in a maths challenge. The table shows their scores for two tests, a mental test and a problem-solving test.

Mental	25	32	32	43	47	50	55	58	61	65	68	72
Problem	32	30	36	40	50	59	52	53	62	60	48	73

Harry did the mental test and scored 53.

How many marks would you expect him to get in the problem-solving test?

AU 4 Describe what you would expect a scatter graph to look like if someone said that it showed no correlation.

12.3 Surveys

HOMEWORK 12C

1 'People like the video rental store to be open 24 hours a day.'

a To see whether this statement is true, design a data collection sheet that will allow you to capture data while standing outside a video rental store.

b Does it matter at which time you collect your data?

2 The youth club wanted to know which types of activities it should plan, e.g. craft, swimming, squash, walking, disco etc.

a Design a data collection sheet that you could use to ask the pupils in your school which activities they would want in a youth club.

b Invent the first 30 entries on the chart.

3 What types of film do your age group watch at the cinema the most? Is it comedy, romance, sci-fi, action, suspense or something else?

   a Design a data collection sheet to be used in a survey of your age group.

   b Invent the first 30 entries on your sheet.

4 Design a two-way table to show the type of music students prefer to listen to in different year groups.

FM 5 Sonia wants to find out who brings packed lunches to school.

She decides to investigate the hypothesis:

'Vegetarians bring packed lunches to school'.

   a Design a data capture form that Sonia could use to help her do this.

   b Sonia records information from a sample of 50 vegetarians and 40 non-vegetarians. She finds that seven vegetarians and two non-vegetarians brought packed lunch to school.

   Based on this sample, is the hypothesis correct?

   Explain your answer.

PS 6 How long do students in your school year spend on homework at the weekend? Design a data collection sheet to assist you find this out.

AU 7 Nusahaa is asked to find out how long people would like the local supermarket to be open for.

She says, 'I will just ask them how long they want it open for.'

What would be the best way for Nusahaa to capture the replies?

HOMEWORK 12D

1 Design a questionnaire to test the following statement.

'Young people aged 15 and under will not tell their parents when they have been given a detention at school, but the over 15s will always let their parents know.'

2 'Boys will use the Internet almost every day but girls will only use it about once a week.' Design a questionnaire to test this statement.

3 Design a questionnaire to test the following hypothesis.

'When you are in your twenties, you watch less TV than any other age group.'

4 While on holiday in Wales, I noticed that in the supermarkets there were a lot more women than men, and the only men I did see were over 65.

   a Write down a hypothesis from the above observation.

   b Design a questionnaire to test your hypothesis.

FM 5 Scott and Gabriel are doing a survey on the type of music 15-year-olds listen to.

a This is one question from Scott's survey.

> 15-year-olds listen to Pop.
>
> Don't you agree?
>
> Strongly agree ☐   Agree ☐   Don't know ☐

Give two criticisms of Scott's question.

b This is a question from Gabriel's survey, which he only asked 15-year-olds.

> What kind of music do you prefer to listen to?
>
> Pop ☐   Rock ☐   Folk ☐   Classical ☐   Other ☐

Give two reasons why this is a good question.

c Make up another good question with responses that could be added to this survey.

PS 6 Design a questionnaire to test the hypothesis:

'People who have "Alexander Lessons" live a healthier life'.

AU 7 Dominic wanted to know how much pocket money students in his school year were getting. He used this questionnaire:

> Question: How much pocket money do you get?
>
> Response: Less than £5 ☐   Less than £15 ☐
>
>   Less than £10 ☐   Less than £20 ☐

Explain why the response section of this questionnaire is poor.

12.4 The data-handling cycle

HOMEWORK 12E

Use the data-handling cycle to describe how you would test each of the following hypotheses. State in each case whether you would use Primary or Secondary data.

1 January is the coldest month of the year.

2 Girls are better than boys at estimating weights.

3 More men go to cricket matches than women.

4 The TV show *Strictly Come Dancing* is watched by more women than men.

5 The older you are the more likely you are to go ballroom dancing.

12.5 Other uses of statistics

HOMEWORK 12F

FM 1 The time series shows washing machine production in the UK from November 2007 to November 2008.

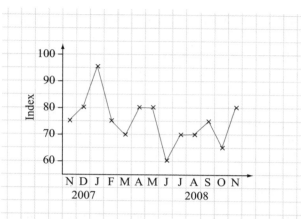

The average production over the first three months shown was 108 000 washing machines.

a Work out an approximate value for the average production over the last three months shown.

b The base month for the index is January 2005, when the index was 100. What was the approximate production in January 2005?

AU 2 The retail price index measures how much the daily cost of living increases or decreases. If 2009 is given a base index number of 100, and 2010 is given 102, what does this mean?

Functional Maths Activity

Election polls

The UK population was faced with a general election in May 2010, and there were daily polls taken to find out how people intended to vote.

The following polls were recorded on 8th April 2010:

Poll	Labour	Conservative	Liberal Democrat	Other
YouGov %	32	37	19	12
ICM %	33	37	21	9
Mori %	30	35	21	14
Comre %	30	37	20	13

Combine all four polls into a single pie chart that represents the opinion of the UK population on 8th April 2010.

13 Probability: Probability of events

13.1 Probability scale

HOMEWORK 13A

1 When throwing a fair dice, state whether each of the following events are impossible, very unlikely, unlikely, even, likely, very likely or certain.
- **a** The score is a factor of 20.
- **b** The score is $3\frac{1}{2}$.
- **c** The score is a number less than six.
- **d** The score is a one.
- **e** The score is a number greater than zero.
- **f** The score is an odd number.
- **g** The score is a multiple of three.

2 Draw a probability scale and put an arrow to show approximately the probability of each of the following events happening.
- **a** It will snow on Christmas Day this year.
- **b** The sun will rise tomorrow morning.
- **c** Someone in your class will have a birthday this month.
- **d** It will rain tomorrow.
- **e** Someone will win the Jackpot in the National Lottery this week.

3 Give an event of your own where you think the probability is:
- **a** impossible   **b** very unlikely   **c** unlikely   **d** evens
- **e** likely   **f** very likely   **g** certain.

PS 4 'What is the probability of rolling dice and getting a total of more than 3?' asked Mr Ball. 'Choose from impossible, unlikely, even, likely, certain.'
Elliott said, 'It must depend on how many dice I use sir.'
Is Elliott correct? Explain your answer.

AU 5 'I have bought 10 raffle tickets this week, so I have a very good chance of winning.'
Explain what might be wrong with this statement.

13.2 Calculating probabilities

HOMEWORK 13B

Example   A bag contains five red balls and three blue balls. A ball is taken out at random.
What is the probability that it is: **a** red   **b** blue   **c** green?

- **a** There are five red balls out of a total of eight, so P(red) = $\frac{5}{8}$.
- **b** There are three blue balls out of a total of eight, so P(blue) = $\frac{3}{8}$.
- **c** There are no green balls, so P(green) = 0.

1 When drawing a card from a well-shuffled pack of cards, what is the probability of each of the following events? Remember to cancel down the probability fraction if possible.
- **a** Drawing an Ace.
- **b** Drawing a picture card.
- **c** Drawing a Diamond.
- **d** Drawing a Queen or a King.
- **e** Drawing the Ace of Spades.
- **f** Drawing a red Jack.
- **g** Drawing a Club or a Heart.

2 The numbers 1 to 10 inclusive are placed in a hat. Irene takes a number out of the hat without looking. What is the probability that she draws:
- **a** the number 10
- **b** an odd number
- **c** a number greater than 4
- **d** a prime number
- **e** a number between 5 and 9?

3 A bag contains two blue balls, three red balls and four green balls. Frank takes a ball from the bag without looking. What is the probability that he takes out:
- **a** a blue ball
- **b** a red ball
- **c** a ball that is not green
- **d** a yellow ball?

4 In a prize raffle there are 50 tickets: 10 coloured red, 10 coloured blue and the rest coloured white. What is the probability that the first ticket drawn out is:
- **a** red
- **b** blue
- **c** white
- **d** red or white
- **e** not blue?

5 A bag contains 15 coloured balls. Three are red, five are blue and the rest are black. Paul takes a ball at random from the bag.
- **a** Find:
  - **i** P(he chooses a red)
  - **ii** P(he chooses a blue)
  - **iii** P(he chooses a black).
- **b** Add together the three probabilities. What do you notice?
- **c** Explain your answer to part **b**.

6 Boris knows that when he plays a game of chess, he has a 65% chance of winning a game and a 15% chance of losing a game. What is the probability that he draws a game?

FM 7 Adam, Emily, Katie, Daniel and Maria are all sitting in Mrs Odell's class during lunch break. Mrs Odell wants two of these students to tidy a cupboard.
- **a** Write down all the possible combinations of two students.
- **b** How many pairs give two girls?
- **c** What is the probability of choosing two girls at random?
- **d** How many pairs give a girl and a boy?
- **e** What is the probability of, at random, choosing a boy and a girl?
- **f** What is the probability of choosing two boys?

PS 8 The following information is known about the following classes at a school.

Year	Y7		Y8		Y9		Y10		Y11	
Class	Boys	Girls	Boys	Girls	Boys	Girls	Boys	Girls	Boys	Girls
Pets	7	8	8	9	10	9	8	9	8	11
No pets	4	5	4	5	6	8	5	6	5	4

A class representative is chosen at random from each class.
Which class has the best chance of choosing a boy who also has a pet as its representative?

AU 9 George is playing a game of tennis.
His sister says, 'He either wins or loses, so he must have an even chance of winning.'
Explain why George's sister might not be correct.

13.3 Probability that an outcome of an event will not happen

HOMEWORK 13C

> **Example**   What is the probability of not picking an Ace from a pack of cards?
>
> First, find the probability of picking an Ace: P (picking an Ace) $= \frac{4}{52} = \frac{1}{13}$
> Therefore, P (not picking an Ace) $= 1 - \frac{1}{13} = \frac{12}{13}$

1 **a** The probability of winning a prize in a tombola is $\frac{1}{25}$. What is the probability of not winning a prize in the tombola?

b The probability that it will rain tomorrow is 65%. What is the probability that it will not rain tomorrow?

c The probability that Josie wins a game of tennis is 0.8. What is the probability that she loses a game?

d The probability of getting a double six when throwing two dice is $\frac{1}{36}$. What is the probability of not getting a double six?

2 Harvinder picks a card from a pack of well-shuffled playing cards. Find the probability that she picks:

a **i** a King   **ii** a card that is not a King

b **i** a Spade   **ii** a card that is not a Spade

c **i** a 9 or a 10   **ii** neither a 9 nor a 10.

3 The following letters are put into a bag.

| A | B | R | A | C | A | D | A | B | R | A |

a Stan takes a letter at random. What is the probability that:
  i he takes a letter A   **ii** he does not take a letter A?

b Pat takes an R and keeps it. Stan now takes a letter from the remaining letters.
  i What is the probability that he takes a letter A?
  ii What is the probability that he does not take a letter A?

FM 4 The starting row in a board game is:

Start	Chelsea Park	Take a Chance	London Road	Carter Road	Pay £500 in tax	Banner Road	Struen Road	Rest area

You roll a dice and move, from the start, the number of places shown by the dice.
What is the probability of **not** landing on:

a A purple square?

b The Pay Tax square?

c A coloured square?

PS 5 Elijah and Harris are playing a board game. Each player rolls one dice to move.
On the next go, Elijah will have to miss a turn if he rolls an odd number. Harris will miss a turn if he rolls a 1 or a 2.
Who has the best chance of not missing a turn on their next go?

AU 6 Arran is told, 'The chance of you losing this game is only 0.1'
Arran says, 'So my chance of winning is 0.9'
Explain why Arran might be wrong.

13.4 Addition rule for events

HOMEWORK 13D

1 Shaheeb throws an ordinary dice. What is the probability that he throws:
a an even number   **b** a 5   **c** an even number or 5?

2 Jane draws a card from a pack of cards. What is the probability that she draws:
a a red card   **b** a black card   **c** a red or a black card?

3 Natalie draws a card from a pack of cards. What is the probability that she draws one of the following?
a Ace   **b** King   **c** Ace or King

4 A letter is chosen at random from the letters in the word STATISTICS. What is the probability that the letter will be:
a an S   **b** a vowel   **c** an S or a vowel?

5 John has a bag containing six red, five blue and four green balls. One ball is picked from the bag at random. What is the probability that the ball is:
a red or blue   **b** not blue   **c** pink   **d** red or not blue?

6 A spinner has numbers and colours on it, as shown in the diagram. Their probabilities are given in the tables.

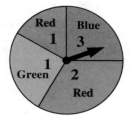

Red	0.5
Green	0.25
Blue	0.25

1	0.4
2	0.35
3	0.25

When the spinner is spun what is the probability of each of the following?
a Red or blue   **b** 2 or $\overset{.}{3}$ **c** 3 or blue   **d** 2 or green
e **i** Explain why the answer to **c** is 0.25 and not 0.5.
     ii What is the answer to P(2 or red)?

7 Debbie has ten CDs in her multi-change CD player, four of which are rock, two are dance and four are classical. She puts the player on random play. What is the probability that the first CD will be:
a rock or dance   **b** rock or classical   **c** not rock?

8 Frank buys a dozen free-range eggs. The farmer tells him that a quarter of the eggs his hens lay have double yolks.
a How many eggs with double yolks can Frank expect to get?
b He cooks three and finds they all have a single yolk. He argues that he now has a 1 in 3 chance of a double yolk from the remaining eggs. Explain why he is wrong.

FM 9 A TV game show has contestants choosing a door to open.
There are 15 doors to choose from, all the same size.
Three doors have small prizes behind them.
Two doors have medium prizes behind them.
One door has the major prize behind it.
All the other doors have 'lose' behind them.
What is the probability that a contestant choosing a door at random does not lose?

PS 10 At a school fair, the cloakroom tickets with numbers from 1 to 100 were used to indicate prizes on a tombola.

All numbers ending with a 0, 2 or a 3 were put on small prizes.

All numbers ending with a 1, 4, 6 or 9 were thrown away.

All numbers ending with a 5 were put on a large prize.

All numbers ending with a 7 or 8 were put on a medium prize.

a How many large prizes were there?

b How many medium prizes were there?

c How many prizes were there altogether?

d What was the probability of winning a large prize?

AU 11 The probability of it snowing on any particular day in January is $\frac{1}{4}$.

Ryan says, 'So the chance of it snowing at all in January is certain, since 31 quarters is more than 1.'

Explain why Ryan is wrong.

13.5 Experimental probability

HOMEWORK 13E

1 Katrina throws two dice and records the number of doubles that she gets after various numbers of throws. The table shows her results.

Number of throws	10	20	30	50	100	200	600
Number of doubles	2	3	6	9	17	35	102

a Calculate the experimental probability of a double at each stage that Katrina recorded her results.

b What do you think the theoretical probability is for the number of doubles when throwing two dice?

2 Mary made a six-sided spinner, like the one shown in the diagram. She used it to play a board game with her friend Jane. The girls thought that the spinner wasn't very fair as it seemed to land on some numbers more than others. They spun the spinner 120 times and recorded the results. The results are shown in the table.

Number spinner lands on	1	2	3	4	5	6
Number of times	22	17	21	18	26	16

a Work out the experimental probability of each number.

b How many times would you expect each number to occur if the spinner is fair?

c Do you think that the spinner is fair? Give a reason for your answer.

3 In a game at the fairground a player rolls a coin onto a squared board with some of the squares coloured blue, green or red. If the coin lands completely within one of the coloured squares the player wins a prize. The table below shows the probabilities of the coin landing completely within a winning colour.

Colour	Blue	Green	Red
Probability	0.3	0.2	0.1

a On one afternoon 300 games were played. How many coins would you expect to land on:　**i** a blue square　**ii** a green square　**iii** a red square?

b What is the probability that a player loses a game?

FM **4** A check was done one week to see how many passengers on the Number 79 bus were pensioners. The table shows the information gathered.

	Mon	**Tue**	**Wed**	**Thu**	**Fri**
Passengers	950	730	1255	796	980
Pensioners	138	121	168	112	143

For each day, what is the probability that a pensioner was the 400th passenger to board the bus that day?

PS **5** Andrew made a six-sided spinner.

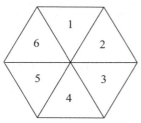

He tested it out to see if it was fair.
He spun the spinner 240 times and recorded the results in a table.

Number spinner lands on	**1**	**2**	**3**	**4**	**5**	**6**
Frequency	43	38	31	41	42	44

Do you think the spinner is fair?
Give reasons for your answer.

AU **6** Aleena tossed a coin 50 times to see how many tails she would get.
She said, 'If this is a fair coin, then I should get 25 tails.'
Explain why she was wrong.

13.6 Combined events

HOMEWORK 13F

1 Copy and complete the sample space diagram to show the total score when two dice are thrown together.

a What is the most likely score?

b Which two scores are least likely?

c Write down the probability of getting a double six.

d What is the probability that a score is:

　i 11

　ii 4

　iii greater than 9

　iv an odd number

　v 4 or less

　vi a multiple of 4?

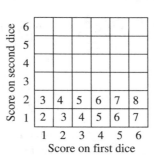

2 Copy and complete the sample space diagram to show the outcomes when a dice and a coin are thrown together.

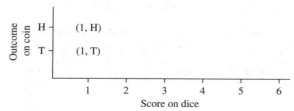

Find the probability of getting:
a a Head and a score of 6
b a Tail and an even score
c a score of 3.

3 Elaine throws a coin and spins a 5-sided spinner that is numbered 1–5. One possible outcome is (Heads, 5).
a List all the possible outcomes.
b What is the probability of getting Tails on the coin and an odd number on the spinner?

4 A bag contains five discs that are numbered 2, 4, 6, 8 and 10. Sharleen takes a disc at random from the bag and puts the disc back. She shakes the bag and takes a disc again. She adds together the two numbers on the discs she has chosen.

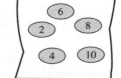

a Copy and complete the table to show all the possible totals.

First number

Second number		2	4	6	8	10
	2					
	4					
	6					
	8					
	10					

b Find the probability that the total is:
i 12 **ii** 20 **iii** 15 **iv** a square number **v** a multiple of 3.

FM 5 Samuel is playing a board game with his sister Ellen. They each roll two normal dice and add the two faces showing.
On their next go, Samuel must roll a total greater than 6 to avoid losing, and Ellen goes to jail if she rolls a double.
a What is the probability of Samuel not losing next go?
b What is the probability of Ellen not going to jail next go?

PS 6 Two eight-sided dice showing the numbers 1 to 8 are thrown at the same time.
What is the probability that the product of the two dice is an odd square number?

AU 7 Isaac rolls two dice and adds both numbers. He wants to know the probability of rolling two dice that will give him a total of a prime number.
Explain why a probability space diagram will help him.

13.7 Expectation

HOMEWORK 13G

1 I throw an ordinary dice 600 times. How many times can I expect to get a score of 1?

2 I toss a coin 500 times. How many times can I expect to get a tail?

3 I draw a card from a pack of cards and replace it. I do this 104 times. How many times would I expect to get:

a a red card **b** a Queen **c** a red seven **d** the Jack of Diamonds?

4 The ball in a roulette wheel can land on any number from 0 to 36. I always bet on the same block of numbers 1–6. If I play all evening and there is a total of 111 spins of the wheel in that time, how many times could I expect to win?

5 I have five tickets for a raffle and I know that the probability of my winning the prize is 0.003. How many tickets were sold altogether in the raffle?

6 In a bag there are 20 balls, ten of which are red, three yellow, and seven blue. A ball is taken out at random and then replaced. This is repeated 200 times. How many times would I expect to get:

a a red ball **b** a yellow or blue ball
c a ball that is not blue **d** a green ball?

7 A sampling bottle contains black and white balls. It is known that the probability of getting a black ball is 0.4. How many white balls would you expect to get in 200 samples if one ball is sampled each time?

8 **a** Fred is about to take his driving test. The chance he passes is $\frac{1}{3}$. His sister says 'Don't worry if you fail because you are sure to pass within three attempts because $3 \times \frac{1}{3} = 1$'. Explain why his sister is wrong.

b If Fred does fail would you expect the chance that he passes next time to increase or decrease? Explain your answer.

9 Kara rolls two dice 200 times.

a How many times would she expect to roll a double?
b How many times would she expect to roll a total score greater than 7?

FM 10 An opinion poll used a sample of 200 voters in one area. 112 said they would vote for Party A. There are a total of 50 000 voters in the area.

a If they all voted, how many would you expect to vote for Party A?
b The poll is accurate within 10%. Can Party A be confident of winning?

PS 11 A roulette wheel has 37 spaces for the ball to land on. The spaces are numbered 0 to 36. I always bet on a prime number.
If I play the game all evening and anticipate playing 100 times, how many times would I expect to win on the roulette table?

AU 12 A headteacher is told that the probability of any student being left-handed is 0.14
How will she find out how many of her students she should expect to be left-handed?

13.8 Two-way tables

HOMEWORK 13H

1 Two dice are thrown together. Draw a probability diagram to show the total score.

a What is the probability of a score that is:
i 7 **ii** 5 or 8 **iii** bigger than 9 **iv** between 2 and 5
v odd **vi** a non-square number?

2 Two dice are thrown. Draw a probability diagram to show the outcomes as a pair of coordinates.

What is the probability that:

a the score is a 'double'

b at least one of the dice shows 3

c the score on one die is three times the score on the other die

d at least one of the dice shows an odd number

e both dice show a 5

f at least one of the dice will show a 5

g exactly one die shows a 5?

3 Two dice are thrown. The score on the first dice is doubled and the score on the second dice is subtracted.

Complete the probability space diagram.

For the event described above, what is the probability of a score of:

a 1

b a negative number

c an even number

d 0 or 1

e a prime number?

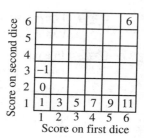

4 When two coins are tossed together, what is the probability of:

a 2 heads or 2 tails **b** a head and a tail **c** at least 1 head?

5 When three coins are tossed together, what is the probability of:

a 3 heads or 3 tails **b** 2 tails and 1 head **c** at least 1 head?

6 When a dice and a coin are thrown together, what is the probability of each of the following outcomes?

a You get a tail on the coin and a 3 on the dice.

b You get a head on the coin and an odd number on the dice.

7 Max buys two bags of bulbs from his local garden centre. Each bag has 4 bulbs. Two bulbs are daffodils, one is a tulip and one is a hyacinth. Max takes one bulb from each bag.

Hyac				HH
Tulip	DT			
Daff				
Daff	DD	DD	TD	
	Daff	**Daff**	**Tulip**	**Hyac**

a There are six possible different pairs of bulbs. List them all.

b Complete the sample space diagram.

c What is the probability of getting two daffodil bulbs?

d Explain why the answer is not $\frac{1}{6}$.

FM 8 A wholesaler sells two varieties of cucumbers – French and British. They are sold in boxes of 5 kg.

The following table shows the mean weight and range of weights for the cucumbers in each box.

	Mean weight	**Range of weights**
British	245 g	60 g
French	356 g	110 g

a James buys a box of cucumbers. He wants all the cucumbers to be about the same weight.

Which box should he buy? Give a reason for your choice.

b Min buys a box of cucumbers. He wants as many cucumbers as possible in the box. Which box should he buy? Give a reason for your choice.

c In the box of British cucumbers, the heaviest cucumber was 270 g. What was the weight of the lightest cucumber?

PS 9 Two fair six-sided spinners are spun:
Spinner A has numbers 1, 5, 8, 9, 10 and 12.
Spinner B has numbers 2, 3, 4, 5, 7 and 9.
What is the probability that, when the two spinners are spun, the two numbers given will multiply to a total greater than 35?

AU 10 Senuri planted some broad bean plants in her greenhouse, while her husband Christos planted some in the garden.
After the summer, they compared their sweet peas.

	Garden	Greenhouse
Mean length of broad bean	17 cm	15.3 cm
Mean weight of broad bean	85 g	94 g

Use the data in the table to explain who had the better crop of broad beans.

Functional Maths Activity

School fete

The following grid was used in a school fete as a 'roll a 10p' game.
A 10p coin was rolled down a chute to roll onto the board, which measured 21 cm by 21 cm.

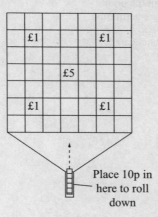

Place 10p in here to roll down

If the 10p landed:

- completely on any square, the player got their 10p back
- completely on a square marked £1, the player won £1
- completely on the square marked £5, the player won £5

1 Conduct an experiment to find out what the probability is of each of these outcomes: a player winning £5, a player winning £1, or a player winning their money back.

2 Use your answers to Question **1** to estimate how much money this game would win or lose if 500 people were to play it.

William Collins' dream of knowledge for all began with the publication of his first book in 1819. A self-educated mill worker, he not only enriched millions of lives, but also founded a flourishing publishing house. Today, staying true to this spirit, Collins books are packed with inspiration, innovation and practical expertise. They place you at the centre of a world of possibility and give you exactly what you need to explore it.

Collins. Freedom to teach.

Published by Collins
An imprint of HarperCollins*Publishers*
77–85 Fulham Palace Road
Hammersmith
London
W6 8JB

Browse the complete Collins catalogue at
www.collinseducation.com

© HarperCollins*Publishers* Limited 2010

10 9 8 7 6 5 4 3 2 1

ISBN-13 978-0-00-733987-7

Brian Speed, Keith Gordon, Keith Evans and Trevor Senior assert their moral rights to be identified as the authors of this work

All rights reserved. No part of this publication may be reproduced, stored in a retrieval system, or transmitted in any form or by any means, electronic, mechanical, photocopying, recording or otherwise, without the prior written permission of the Publisher or a licence permitting restricted copying in the United Kingdom issued by the Copyright Licensing Agency Ltd., 90 Tottenham Court Road, London W1T 4LP.

British Library Cataloguing in Publication Data
A Catalogue record for this publication is available from the British Library

Commissioned by Katie Sergeant
Project managed by Patricia Briggs
Edited by Brian Asbury
Cover design by Angela English
Concept design by Nigel Jordan
Illustrations by Wearset Publishing Services
Typesetting by Wearset Publishing Services
Production by Simon Moore
Printed and bound by L.E.G.O. S.p.A. Italy

Acknowledgement
With thanks to Chris Pearce (Teaching and Learning Advisor, North Somerset).

Important information about the Student Book CD-ROM
The accompanying CD-ROM is for home use only. You cannot copy or save the files to your hard drive and it will work only when placed in the CD-ROM drive.